MINISTRY OF THE HOLY SPIRIT IN A TECHNOLOGICAL AGE

Bishop Kevin E. Knight, Sr., Th.D.

BK Royston Publishing
http://www.bkroystonpublishing.com

Cover Design: R. J. Smith

ISBN-13: 978-1-971868-71-4

Printed in the United States of America

DEDICATION

This book is dedicated, first and foremost, to God's faithfulness, and the Spirit has guided, corrected, sustained, and empowered me through every season of my life and ministry.

I acknowledge and honor the spiritual foundation my parents, Bishop John Knight and Mother Fannie Knight, laid, whose lives modeled reverence for God, obedience to the Holy Spirit, and unwavering commitment to the Church. What they planted in prayer, discipline, and example continues to bear fruit.

The technological age has not altered the truth of Ruach. The church may design programs and strategies, but unless the breath of God animates them, they remain lifeless structures.

ACKNOWLEDGEMENT

To my beloved wife, **Mary L. Knight**, my covenant partner of forty-eight-plus years—thank you for your steadfast love, patience, sacrifice, and unshakable support. This journey would not have been possible without your faith, strength, and prayerful covering.

To my son Kevin Jr., my son-in-love, Wil Young, and most especially to my daughter, Shantae' Merie—thank you for your unwavering support of the ministry. Shantae', your devotion, your encouragement, and your readiness to stand with purpose in every season have strengthened both me and the work of the Lord.

I am deeply grateful for my church family, *Heavenly Temple Church of God in Christ,* whose prayers, faithfulness, and shared labor in the gospel have been a constant source of strength and joy. You are not merely a congregation, but a spiritual family.

I also acknowledge the many leaders, mentors, colleagues, and friends God has placed in my life over the years. Each conversation, correction, encouragement, and shared labor has shaped this work in ways seen and unseen.

This book is the fruit of many hands, many prayers, and many years — all under the gracious leading of the Holy Spirit.

A pastor may have access to trends, metrics, surveys, and demographic profiles, yet still fail to hear what the Spirit is saying to the church.

FOREWORD

We live in a time that seriously challenges the credibility and reverence of the ancient Scriptures. Yet the Church is still called and commissioned to proclaim the centrality of the Gospel of Jesus Christ. As Jesus promised, *"But the Helper, the Holy Spirit, whom the Father will send in My name, He will teach you all things and bring to your remembrance all things that I have said to you"* (John 14:26).

In these pages, the author draws us back to that promise, calling us to live and lead with Spirit-empowered discernment in an age of AI and accelerating technology. With devotion to Scripture and a life committed to Gospel ministry, he also sounds a needed caution: we must never substitute Ruach—the Spirit and breath of God—for AI or any technology.

While the author acknowledges the value technology can bring to administering some church-related functions, he also sounds the alarm about the negative influence AI and technology can have on individuals and congregations when they are misperceived or misused. I encourage clergy, laity, and leaders to read this book with reverent openness—seeking insight for life and living according to the Scriptures as the Holy Spirit leads.

The "noise" and "information overload" of our day create a constant battle for our attention—and that is exactly why this book is needed. As some in the Church embrace technology for ministry, we must stay focused on what truly matters. With the proliferation of "fake news" and manufactured narratives, the Church—under the operation of the Holy Spirit—must discern truth from lies and deception, any of which can lead people and congregations astray. We must not forget the Church's origin or the impact of the Holy Spirit at Pentecost. The Holy Spirit empowered the Church amid chaos and persecution, and He still does.

"AI can provide a 'smart' answer, but it cannot provide a 'wise' one, because wisdom requires a soul, a conscience, and a relationship with the Creator."

The author acknowledges that we can embrace technology and automation, but we must never give them the place of prominence that belongs to the Holy Spirit. The Spirit is our source and guide, and He strengthens every area of our growth—today and in the days ahead. We must continue to trust and obey His leading, as He empowers believers with boldness and anointing for ministry.

The Church today—and in the future—must be led by men and women of character: authentic, committed believers in the centrality of Jesus Christ, who depend upon the leadership and direction of the Holy Spirit.

Technology can help the Church see what is happening in other regions and sectors of society, but it cannot produce the inward transformation that the Holy Spirit ignites.

The author reminds us that we must not allow our inventions to take the place of our God and Creator. There are many takeaways from this book—let me leave you with a few that especially stood out to me.

Power through prayer and surrender to the Holy Spirit will bring expansion. **"Yet revival born of the Spirit carries fruit that no platform can manufacture."**

"The Spirit moves powerfully where people recognize their dependence."

Finally, **"The cycle is not mechanical but relational. The lesson for the modern church is clear: hunger must be cultivated."**

My prayer is that this book will deepen your hunger for God, sharpen your discernment, and renew your dependence on the Holy Spirit as you follow Jesus Christ.

Blessings.
George S. Patrick

People can treat the Holy Spirit as though He exists to improve ministry efficiency, increase productivity, or intensify religious experience on demand. But the Spirit is not a function within the divine life; He is God. He is not activated like a feature; He is respected, obeyed, and yielded to.

TABLE OF CONTENTS

PREFACE

We are living in a period of rapid acceleration. Technological systems now influence how societies communicate, organize, and make decisions.

Artificial intelligence, once a distant concept, is now a familiar presence in everyday life, shaping processes that once relied on sustained human judgment. Efficiency has increased. Access has expanded. Speed has become expected.

In this environment, the church faces a defining responsibility. It must engage the realities of the modern world without allowing those realities to redefine spiritual authority, spiritual formation, or spiritual dependence.

The question before the church is not whether technology will be used, but whether its use will remain governed by discernment, prayer, and faithfulness to God.

This book is written from theological conviction and pastoral concern rather than cultural anxiety. It does not treat innovation as an inherent threat, nor does it assume that technological advancement automatically serves God's purposes. Tools themselves are morally neutral. The deeper issue is what the church trusts, prioritizes, and allows to shape its understanding of power, wisdom, and effectiveness.

Accordingly, this work is not a technical analysis of artificial intelligence, nor is it an ethical manual addressing every emerging digital dilemma. Those discussions have their place, but this book focuses on something more foundational.

This book examines the ministry of the Holy Spirit as the church navigates an age increasingly defined by automation, optimization, and *control*. It asks whether spiritual dependence can be sustained when efficiency is rewarded, and waiting is discouraged.

The chapters that follow move intentionally from biblical foundations to historical perspective and contemporary application. Scripture establishes continuity.

History provides context. Present-day realities offer a testing ground. Throughout, the goal is not reaction but discernment—not alarm, but alignment.

The concern is not whether the church can function effectively in a technological age, but whether it can do so without surrendering what cannot be replicated, automated, or engineered.

This book is best approached slowly and prayerfully. It is not written to provoke controversy or to win arguments, but to encourage sober reflection. Readers are invited to examine not only ministry practices but the assumptions beneath them.

Modern tools may assist the church's mission, but they must never become substitutes for spiritual authority or dependence upon God.

The Holy Spirit has not been withdrawn from the life of the church. He has not diminished in relevance or power.

The same Spirit who has sustained the people of God across generations remains present and active today. The challenge of the present age is not His absence, but our attentiveness.

This book will help leaders, ministers, and believers stay clear-minded, spiritually grounded, and faithfully dependent as they serve in an increasingly complex world.

If it contributes to renewed reverence, deeper prayer, and greater discernment, then it will have fulfilled its purpose.

The church must maintain a clear theological distinction: AI can organize information, but it cannot produce revelation. It can summarize a doctrine, but it cannot anoint it. It can mimic the tone of devotion, but it cannot produce the fruit of devotion.

WHY THE HOLY SPIRIT MUST BE RECLAIMED IN THE MODERN CHURCH

We are living in a period of rapid acceleration. Technological systems now influence how societies communicate, organize, and make decisions. Artificial intelligence, once a distant concept, is now a familiar presence in everyday life, shaping processes that once relied on sustained human judgment. Efficiency has increased. Access has expanded. Speed has become expected.

In this environment, the church faces a defining responsibility. It must engage the realities of the modern world without allowing those realities to redefine spiritual authority, spiritual formation, or spiritual dependence. The question before the church is not whether technology will be used, but whether its use will remain governed by discernment, prayer, and faithfulness to God.

Who This Book Is For

This work is designed for pastors, church leaders, and theologically literate laypeople who are navigating the tension between digital innovation and biblical fidelity. It serves as a guide for those who recognize that while our tools change, our Source remains immutable.

This book is written from theological conviction and pastoral concern rather than cultural anxiety. It does not treat innovation as an inherent threat, nor does it assume that technological advancement automatically serves God's purposes. Tools, in themselves, are morally neutral. The deeper issue is posture—what the church trusts, prioritizes, and allows to shape its understanding of power, wisdom, and effectiveness.

Accordingly, this work is not a technical analysis of artificial intelligence, nor is it an ethical manual addressing every emerging digital dilemma. Those discussions have their place, but this book focuses on something more foundational.

This book examines the ministry of the Holy Spirit as the church navigates an age increasingly defined by automation, optimization, and control. It asks whether spiritual dependence can endure when efficiency is rewarded, and waiting is discouraged.

The chapters that follow move intentionally from biblical foundations to historical perspective and contemporary application. Scripture establishes continuity. History provides context. Present-day realities offer a testing ground.

Throughout, the goal is not reaction but discernment; not alarm, but alignment. The concern is not whether the church can function effectively in a technological age, but whether it can do so without surrendering what cannot be replicated, automated, or engineered.

This book is best approached slowly and prayerfully. It is not written to provoke controversy or to win arguments, but to encourage sober reflection. Readers are invited to examine not only ministry practices, but the assumptions beneath them.

Modern tools may assist the church's mission, but they must never replace spiritual authority or dependence on God.

The Holy Spirit has not been withdrawn from the life of the church. He has not diminished in relevance or power. The same Spirit who has sustained the people of God across generations remains present and active today. The challenge of the present age is not His absence, but our attentiveness.

This book will help leaders, ministers, and believers remain clear-minded, spiritually grounded, and faithfully dependent as they serve in an increasingly complex world. If it contributes to renewed reverence, deeper prayer, and greater discernment, then it will have fulfilled its purpose.

THE SPIRIT OF GOD FROM CREATION TO NOW

Foundational Truth for Discernment in a Technological Age

THE BREATH OF GOD: UNDERSTANDING RUACH

The first mention of the Spirit of God appearing before light, which is spoken into existence, establishes a cosmic reality that transcends all human innovation. In an age fascinated with intelligence, speed, automation, and digital mastery, Scripture begins by reminding us that life does not originate in machinery, systems, or human advancement. While servers store vast amounts of information and networks transmit data at lightning speed, they cannot generate spiritual life.

From Scripture's opening lines, the Spirit is presented as *Ruach,* the active, animating breath of God. Genesis records that "the earth was without form, and void; and darkness was upon the face of the deep. And the Spirit of God moved upon the face of the waters" (**Genesis 1:2**). This movement is not decorative.

It signals divine involvement in ordering chaos before any structure is established.

Ruach means breath, wind, or spirit, describing a movement that cannot be seen yet can be felt. Wind needs no permission to move, and breath does not wait for instructions to sustain life. He is not introduced as an abstract idea but as God's animating presence at work before creation takes shape. As the Psalmist declares, "By the word of the Lord were the heavens made; and all the host of them by the breath of his mouth" (**Psalm 33:6**).

Humanity does not generate its own life; the divine breath imparts it.

This same Spirit who hovered over the waters is the one who parted seas, empowered prophets, and raised Christ from the dead. The concept of breath also communicates essential vitality, as breath is inseparable from life. When Genesis describes God forming man from the dust and breathing into his nostrils the breath of life (**Genesis 2:7**), it emphasizes that we are recipients of life rather than its creators. Every heartbeat traces back to this divine breath, and every living soul exists because the Spirit sustains it (**Job 33:4**). He moves over emptiness to bring fullness, and over darkness to bring illumination.

Understanding this corrects the modern misunderstanding that the Spirit is merely a religious atmosphere or an emotional intensity during worship. Furthermore, the Spirit's activity in creation sets the stage for His role in redemption. The same power that brought order to primordial chaos is the power that brings order to a fractured heart.

As noted by theologians like **Gordon Fee**, the Spirit's role in the New Covenant is the fulfillment of this life-giving presence, moving from external to internal indwelling in the believer. He is not confined to the sanctuary or restricted to ritual; He is the living breath of God operating across all of material reality.

Without *Ruach*, there is no life, no church, and no transformation. We must see His work as a continuous thread from the beginning of time into our present moment. This opening truth is essential for discernment in a technological age. If the Spirit is first introduced to us not as an accessory to religion but as the source of life itself, then everything that follows must be measured by that reality. Discernment begins when the church remembers that information is not the same as illumination, output is not the same as life, and movement on a screen is not the same as the movement of God.

- *ANCIENT LIFE IN A DIGITAL WORLD*

The technological age has not altered the truth of *Ruach*. The church may design programs and strategies, but unless the breath of God animates them, they remain lifeless structures. In a digital civilization that often compartmentalizes faith, we must remember that the Spirit governs both creation and conscience. Modern tools may assist the church's mission, but they must never replace spiritual authority or dependence on God.

Recognizing this cosmic role keeps believers from reducing the Spirit to mere church activity and builds confidence that Digital networks do not threaten him. If the Spirit orders galaxies and sustains life, He is certainly not intimidated by artificial intelligence or the speed of global communication. The same Spirit who hovered over primordial waters oversees modern innovation, and nothing progresses beyond His authority.

Even the documents of **Vatican II**, such as *Gaudium et Spes*, remind us that while human progress is a sign of God's greatness, it must be governed by the Spirit to serve the true good of humanity. This truth should humble the modern church. We are often tempted to believe that because our tools are more advanced, our insight is deeper; because our communication is faster, our mission is stronger; because our content is more polished, our witness is more effective. Yet Scripture teaches the opposite kind of dependence. The church has never been sustained by novelty; God's presence has always sustained it. What is ancient in origin is still superior in power.

When we rely on systems rather than the Spirit, the church may appear busy yet slowly become powerless. The Spirit remains the difference between motion and movement; motion can be mechanical and automated, but movement is alive and organic. Technology measures visibility and reach, but it cannot measure the spiritual transformation that only the breath of God provides.

Finally, we must distinguish between the efficiency of our tools and the life of our mission. Machines can replicate the tone of devotion, but they cannot produce its fruit. As we serve in an increasingly complex world, the challenge is not the Spirit's absence, but our own attentiveness to His move.

Our task is not to manufacture a movement, but to align ourselves with the one the Spirit has already begun. The church does not need to fear technology itself; the greater danger is forgetfulness. When believers forget that the Holy Spirit is the source of life, they begin to trust the measurable more than the spiritual. They begin to prize speed over surrender, scale over sanctification, and influence over intimacy with God. Ancient life is still the answer in a digital world because what the church most desperately needs cannot be downloaded, automated, or engineered.

1

THE SPIRIT OF GOD FROM CREATION TO NOW

Foundational Truth for Discernment in a Technological Age

The first mention of the Spirit of God appearing before light, which is spoken into existence, establishes a cosmic reality that transcends all human innovation. In an age fascinated with intelligence, speed, automation, and digital mastery, Scripture begins by reminding us that life does not originate in machinery, systems, or human advancement. While servers store vast amounts of information and networks transmit data at lightning speed, they cannot generate spiritual life.

From the opening lines of Scripture, the Spirit is presented as Ruach, the active, animating breath of God. Genesis records that the earth was without form and void, that darkness covered the deep, and that the Spirit of God moved upon the face of the waters (Genesis 1:2). This movement is not merely decorative; it signals divine involvement in ordering chaos before any structure is established.

Ruach carries the meaning of breath, wind, or spirit, describing a movement that cannot be seen, yet can be felt. Wind does not need permission to move, and breath does not wait for instruction to sustain life. He is not introduced as an abstract idea but as the animating presence of God at work before creation takes shape. Humanity does not generate its own life; life is imparted through the divine breath. This same Spirit who hovered over the waters is the one who parted seas, empowered prophets, and raised Christ from the dead.

The concept of breath also communicates essential vitality, as breath is inseparable from life. When Genesis describes God forming man from the dust and breathing into his nostrils (Genesis 2:7), it emphasizes that we are recipients of life rather than creators of it. Every heartbeat traces back to this divine breath, and every living soul exists because the Spirit sustains it (Job 33:4). He moves over emptiness to bring fullness, and over darkness to bring illumination.

Understanding this corrects the modern misunderstanding that the Spirit is merely a religious atmosphere or an emotional intensity during worship. As noted by theologians such as Gordon Fee, who emphasized the Spirit's role in Pauline pneumatology, the Spirit is the "down payment" of the future kingdom, a personal presence rather than a vague influence. Furthermore, the Spirit's activity in creation sets the stage for His role in redemption. The same power that brought order to primordial chaos is the power that brings order to a fractured heart. He is not confined to the sanctuary or restricted to ritual; He is the living breath of God operating across all of material reality.

Without Ruach, there is no life, no church, and no transformation. We must see His work as a continuous thread from the beginning of time into our present moment. This opening truth is essential for discernment in a technological age. If the Spirit is first introduced to us not as an accessory to religion but as the source of life itself, then everything that follows must be measured by that reality. Discernment begins when the church remembers that information is not the same as illumination, output is not the same as life, and movement on a screen is not the same as the movement of God.

ANCIENT LIFE IN A DIGITAL WORLD

The technological age has not altered the truth of Ruach. The church may design programs and strategies, but unless the breath of God animates them, they remain lifeless structures. In a digital civilization that often compartmentalizes faith, we must remember that the Spirit governs both creation and conscience. Modern tools may assist the church's mission, but they must never replace spiritual authority or dependence on God.

Recognizing this cosmic role keeps believers from reducing the Spirit to mere church activity and builds confidence that Digital networks do not threaten him. If the Spirit orders galaxies and sustains life, He is certainly not intimidated by artificial intelligence or the speed of global communication. The same Spirit who hovered over primordial waters oversees modern innovation, and nothing progresses beyond His authority.

This truth should humble the modern church. We are often tempted to believe that because our tools are more advanced, our insight is deeper. Because our communication is faster, our mission is stronger. Because our content is more polished, our witness is more effective. Yet Scripture teaches the opposite kind of dependence. The church has never been sustained by novelty. The presence of God has always sustained it. What is ancient in origin is still superior in power.

When we rely on systems instead of the Spirit, the church may appear busy while slowly becoming powerless. The Spirit remains the difference between motion and movement; motion can be mechanical and automated, but movement is alive and organic. Technology measures visibility and reach, but it cannot measure the spiritual transformation that only the breath of God provides.

Finally, we must distinguish between the efficiency of our tools and the life of our mission. Machines can replicate the tone of devotion, but they cannot produce its fruit. As we serve in a world of increasing complexity, the challenge is not the Spirit's absence, but our own attentiveness to His move. Our task is not to manufacture a movement, but to align ourselves with the one the Spirit has already begun.

The church does not need to fear technology itself. The

greater danger is forgetfulness. When believers forget that the Holy Spirit is the source of life, they begin to trust what is measurable more than what is spiritual. They begin to prize speed over surrender, scale over sanctification, and influence over intimacy with God. Ancient life is still the answer in a digital world because what the church most desperately needs cannot be downloaded, automated, or engineered.

GOD IN THE WORLD, THE CHURCH, AND THE SOUL

The Old Testament presents the Spirit operating in three primary dimensions that provide a roadmap for modern believers.

The World: First, God is seen in the world through creation and providence, where the Spirit orders chaos, sustains life, and directs history.

The Church: Second, God appears in the "theocratic dimension" of the covenant community, where the Spirit empowers leaders and artisans to accomplish divine purposes.

The Soul: Third, God works within the individual soul to transform character, as seen in David's plea for a renewed heart (Psalm 51:10-12).

The "God in the world" dimension reminds us that history is not random. The Spirit is active within material reality, governing natural processes and restraining evil. The Psalms declare that when God sends forth His Spirit, they are created, and the face of the earth is renewed

(Psalm 104:30). This affirms that God's presence is omnipresent, reaching every corner of existence, including the digital spaces we inhabit today.

In the covenant community, the Spirit distributes gifts not for private elevation, but for the fulfillment of God's plan. Whether it was Bezalel receiving wisdom for craftsmanship (Exodus 31:3) or Gideon receiving power for leadership, the Spirit equipped them for the sake of the people. This highlights that spiritual authority flows from divine enablement rather than personal ambition. Without this Spirit-led direction, structure becomes rigid and bureaucratic.

The Spirit's individual work is perhaps the most intimate, moving from external empowerment to internal renovation. Prophetic writings like those of Ezekiel anticipate a time when the heart of stone is replaced with a heart of flesh (Ezekiel 36:26). This inward work reshapes desire and alters motivation, making holiness a natural fruit rather than a forced compliance.

Without this renovation, any growth we see in the church remains purely superficial.

These three dimensions also protect the church from reductionism. Scripture holds these dimensions together: the Spirit is active in the world, present in the covenant community, and at work in the soul. That breadth matters now more than ever because technological culture constantly tempts us to separate what God joins together. If the Spirit governs the world, then He is not absent from public life, cultural upheaval, or technological change. If He empowers the church, then ministry cannot be reduced to branding, organization, or managerial excellence. If He transforms the soul, then no amount of external polish can substitute for inward holiness. Discernment requires us to see all three dimensions clearly, because confusion in any one of them will distort how we interpret the times.

DIVINE PROVIDENCE AND MODERN SYSTEMS

These three dimensions remain intact today, despite our transition into an automated world. Modern technology has not displaced divine providence, and shifting cultural norms have not limited His authority. The Spirit who renews the earth also renews the heart; the Spirit who empowers leaders also disciplines souls.

We must resist the temptation to focus only on personal spirituality while neglecting the Spirit's global mission. When believers understand these dimensions, confusion regarding our role in the modern world decreases. Data can inform our behavior, but it cannot regenerate our desires. Self-help strategies might modify our habits, but they cannot cleanse the conscience. Only the Spirit performs the inward renovation that makes a person a true witness for Christ.

Furthermore, we must recognize that the Spirit's activity is continuous, not seasonal. Just as He once instructed Israel, He continues to function as our teacher and guide.

In the 21st century, insight is not merely a matter of having the most information; it is a matter of spiritual illumination. This allows the church to move with spiritual intelligence through environments that often resist the truth.

We live in systems designed to predict behavior, organize information, and optimize outcomes. Those systems can be useful, but they are limited by design. They can process patterns, but they cannot perceive purpose. They can identify preferences, but they cannot discern conviction. Providence is deeper than prediction; divine guidance is wiser than data modeling.

This becomes especially important in leadership. A pastor may have access to trends, metrics, surveys, and demographic profiles, yet still fail to hear what the Spirit is saying to the church. A ministry may look informed while remaining spiritually dull. A congregation may become highly organized while quietly losing its hunger for God. Providence reminds us that the future of the church is not secured by superior analysis alone, but by faithful submission to the Spirit who sees the end from the beginning.

Finally, the church must remember that its true strategic advantage is not advanced tooling, but the indwelling Holy Spirit. While we have access to sophisticated analytics and demographic data, these can never replace spiritual dependence. If we confuse automated output with divine empowerment, we risk becoming more efficient while simultaneously becoming less alive.

THE HOLY SPIRIT AS A DIVINE PERSON, NOT AN ABSTRACT FORCE

Scripture attributes personal characteristics to the Holy Spirit that move Him beyond the realm of an impersonal energy. He speaks, He instructs, He grieves, and He intercedes. He possesses a mind and searches the deep things of God, implying a conscious awareness and personhood. He does not operate mechanically; He acts intentionally, distributing gifts according to His will rather than human agendas.

Because the Spirit is a person, He can be resisted or grieved (Ephesians 4:30). These descriptions eliminate the possibility that He is merely a symbolic force or a religious atmosphere. Paul appeals to believers through the "love of the Spirit," highlighting the Spirit's capacity for relational intimacy. He dwells with believers and communes with them, and communion fundamentally requires a relationship between people.

Misunderstanding the Spirit as an abstract force leads to a dangerous misuse of His power. If we view Him only as "energy," we might try to harness or direct Him for our own purposes. If we see Him as a mere "influence," we might try to manufacture His presence through emotional triggers. Recognizing His personhood restores a necessary sense of reverence and surrender.

In a world that prizes the ability to manipulate and

control our environment, recognizing the Spirit's personhood is a radical act. One does not command a divine person; one yields to Him. This realization shifts our posture from "using" God to being used by Him. It reminds the church that our primary task is not to harness power, but to maintain a relationship of obedience. This matters even more in the technological age, as modern culture trains us to expect on-demand responsiveness.

We click, prompt, search, request, and receive. That pattern can subtly distort the spiritual life if we are not careful. The Holy Spirit is not a function to be activated on command. He is not a religious interface that exists to satisfy our preferred outcomes. He is Lord. He leads. He convicts. He directs according to divine wisdom, not human convenience.

Once the church loses sight of His personhood, it becomes vulnerable to imitation. Emotional manipulation can be mistaken for anointing. Predictable religious scripting can be mistaken for divine movement. Digital polish can be mistaken for spiritual authority. But when believers remember that the Spirit is a divine person, reverence returns. Prayer deepens. Listening becomes essential. Obedience becomes central again.

THE LIMITS OF AUTOMATION

This understanding protects the church in a technological age where we are tempted to confuse automation with authority. Technology operates through systems, algorithms, and predictable inputs. The Spirit, however, does not function through automation; He responds to surrender, prayer, and faith. He cannot be programmed, engineered, or replicated digitally.

While artificial intelligence can organize information and summarize doctrines, it cannot produce revelation. It can replicate the tone of devotion, but it cannot produce the fruit of a transformed life. The church must be careful not to treat instant digital outputs as spiritual direction. When we trade intimacy for immediacy, we lose the very guidance the Spirit was sent to provide.

Furthermore, the Spirit's guidance in our decisions is often relational and progressive, rather than instant and data-driven. He leads according to divine wisdom, which may at times conflict with the logic of platform algorithms or digital trend cycles. Faithfulness in a technological age might require us to remain silent or deep when the digital world demands we be loud and brief.

Automation helps with repetition, delivery, and organization. It can help us manage communication, streamline workflows, and increase operational consistency. But spiritual leadership is not merely about operational consistency. It requires discernment, conviction, holiness, and sensitivity to the voice of God. Those things do not emerge from code; they emerge from communion.

The danger is not that the church will use tools. The

danger is that the church may begin expecting tools to do what only the Spirit can do. A machine can help distribute a sermon, but it cannot carry the burden that births one. A platform can amplify a message, but it cannot supply the fire behind it. A system can track engagement, but it cannot discern whether hearts are genuinely turning toward Christ.

That is why the church must be vigilant. Efficiency is seductive because it offers visible results. The Spirit often works more deeply than visibly, shaping what cannot be measured in a dashboard. Ultimately, the church must fear self-reliance more than it fears any technological advancement. Technology can extend our reach, but only the Spirit can extend eternal life. If we begin to treat our tools as masters, we subtly shift away from the Spirit's leadership. We must keep our tools as servants and make prayer our primary priority.

The issue is not whether the church will live in a technological age; it already does. The issue is whether the church will remain spiritually governed within it. Chapter One clearly establishes the foundation: the Holy Spirit is not a decorative doctrine in religious language. He is the breath of God, the Lord of life, the divine person who orders chaos, empowers the church, renews the soul, and guides the people of God with wisdom no system can reproduce. Any serious conversation about ministry, discernment, and faithfulness in this age must begin there.

2

GOD'S GOOD PLAN AND THE EMPOWERED CHURCH

Divine Purpose and Spiritual Authority in an Automated World

The story of the Holy Spirit cannot be separated from the story of God's intention for humanity. From the opening chapter of Genesis, God's posture toward humanity is not restrictive but generous. Humanity is created in the image of God and given dominion, stewardship, and purpose. The command to be fruitful and multiply is not merely biological; it is vocational and spiritual (Genesis 1:28). God places humanity within creation not as slaves of chaos but as representatives of divine order. This foundational truth shapes everything that follows: the Spirit's work is not a random intervention. It is the consistent advancement of God's good design.

This matters deeply in an age that increasingly imagines human progress as self-generated. The modern world celebrates acceleration, control, prediction, and optimization. It often assumes that with enough data, automation, and technological refinement, humanity can secure its own future. Scripture challenges that illusion at the root. Human beings do not define their own purpose by innovation alone; they receive purpose from God. The Spirit does not assist humanity in building a self-sufficient world detached from its Maker. He continually calls creation back into alignment with the

Creator's intention.

When sin enters the narrative, God does not abandon His purpose; He initiates redemption. The covenant with Abraham, the deliverance through Moses, the establishment of Israel, and the prophetic promises all testify that God's plan remains intact. Even divine judgment serves redemptive ends. The Spirit works within history to preserve a people through whom blessing will extend to the nations. God's plan has always been outward-facing; the covenant was never intended to be confined to one ethnic group. It was meant to channel global restoration.

Jeremiah records the Lord declaring, "For I know the plans I have for you, plans to prosper you and not to harm you, plans to give you hope and a future" (Jeremiah 29:11). That promise was spoken to exiles, not elites. It reveals that God's purposes operate even in seasons of displacement. The Spirit sustains hope when circumstances appear contradictory. Divine benevolence does not mean the absence of hardship; it means redemptive direction. The Spirit carries forward God's purposes even through exile, disruption, and transition.

That truth speaks powerfully to the present moment. The church now ministers in a world marked by rapid transition, fractured attention, institutional distrust, and technological disruption. Many people feel spiritually displaced even while remaining socially connected. They have information but little peace, access but little anchor, visibility but little meaning. The Spirit still ministers in exile. He still steadies hearts in unstable times. He still reminds the people of God that divine purpose is not erased by upheaval. God's plan is not fragile; it does not collapse under cultural pressure or disappear in an age of automation.

In the New Testament, the scope of God's benevolence becomes unmistakable: "For God so loved the world" (John 3:16). Not a region; not a tribe; the world. Sending the Son demonstrates the seriousness of divine love. The sending of the Spirit ensures that love continues in active power. Redemption is not a theory; it is a reality enacted. The Spirit applies what Christ accomplished. Without the Spirit, the cross remains a historical fact; with the Spirit, it becomes a present transformation.

The technological age does not nullify this benevolent plan. It may amplify reach, but it does not create redemption. The Spirit who once moved through prophets now moves through believers connected across continents. The gospel can travel in seconds. Yet the mechanism of transmission remains secondary. The Spirit still awakens faith; He still softens hearts; He still draws people toward Christ. God's good plan continues, not because of technology, but through it when surrendered to divine purpose. The church must remember the order: God's plan comes first; human tools come later. Mission is not born from machinery; it is born from divine love. The Spirit does not take His cues from innovation; He governs how innovation should be used. Where this order is reversed, ministry becomes clever but shallow. Where this order is honored, even ordinary tools can become vessels for extraordinary grace.

THE TRINITY AND THE CO-EQUAL MINISTRY OF THE SPIRIT

The Holy Spirit is not an assistant to the Father and the Son. He is fully divine, co-equal, and co-eternal. Christian faith is grounded in Trinitarian reality: the Father plans, the Son accomplishes, and the Spirit applies. These are not separate gods performing independent tasks; they are unified persons operating in perfect harmony. Reducing the Spirit's ministry would distort the nature of God.

Scripture affirms this unity repeatedly. At Jesus' baptism, the Son stands in the water, the Spirit descends like a dove, and the Father speaks from heaven (Matthew 3:16-17). The scene is not symbolic theater; it reveals relational distinction within divine unity. Later, Jesus promises that the Father will send the Spirit in His name (John 14:26). The Spirit will glorify the Son and reveal what belongs to the Father (John 16:14-15). Mutual glorification defines the Trinity; none compete, none diminish, and each reveals the other.

As scholar Gordon Fee notes in his work on Pauline pneumatology, the Spirit is the "personal presence of God" himself, not a vague surrogate. The early church experienced this reality. Pentecost did not replace Christ's work but continued it. Jesus ascended bodily, but the Spirit descended personally. Power shifted from a localized presence to an indwelling presence (Acts 1:8). Christ's physical ministry was geographically confined; the Spirit's indwelling ministry multiplies globally. This was not divine improvisation; it was divine design.

The Spirit's equality matters practically. If He is merely an influence, believers may ignore Him; if He is fully God, surrender becomes essential. The Spirit speaks with divine authority, guides with divine wisdom, convicts with divine precision, and comforts with divine intimacy. His ministry is not an optional enhancement; it is an indispensable reality.

This becomes especially important in an age of automation, as it subtly trains people to think in terms of functions rather than persons. Modern systems are valued for speed, output, responsiveness, and utility. That mindset can creep into spiritual life if the church is not careful. The Holy Spirit can slowly be treated as though He exists to improve ministry efficiency, increase productivity, or intensify religious experience on demand. But the Spirit is not a function within the divine life; He is God. He is not activated like a feature; He is adored, obeyed, and yielded to.

When the church loses sight of the Spirit's full divinity, it also loses its sense of holy dependence. It may continue speaking about power while functioning as though power comes from branding, systems, momentum, or organizational precision. It may continue to mention the Trinity in doctrine while operating in practice as though the Father planned salvation, the Son secured salvation, and the church now manages the rest. That is not biblical ministry. The Spirit remains the living executor of God's will in the church and in the world.

In modern church practice, imbalance often emerges. Some emphasize Christ while neglecting the Spirit; others pursue spiritual experiences disconnected from Christ's lordship. Both distortions fracture Trinitarian harmony. The Spirit does not compete with Christ; He magnifies Him. Authentic spiritual empowerment will always elevate Christ, deepen reverence for the Father, and produce obedience. That harmony safeguards the church in a distracted age. If a ministry claims to be Spirit-led but minimizes Christ, it has drifted. If it claims devotion to Christ but resists the Spirit's active governance, it has drifted as well. The Father, Son, and Holy Spirit do not work at cross-purposes. A healthy church, especially in a technologically saturated culture, must keep that Trinitarian order intact. It must not let convenience replace reverence or utility replace surrender.

FROM LAW TO LOVE: THE SHIFT THROUGH CHRIST

The Mosaic law revealed God's holiness. It established boundaries and exposed sin. Yet the law could not transform the human heart; it prescribed righteousness but could not produce it. Punishments were severe because sin was serious. The system preserved a nation but did not regenerate souls. The Spirit was active in this period, but His indwelling presence was not yet universal.

Jesus redefined the fulfillment of the law. He summarized it in two commands: love God and love your neighbor (Matthew 22:37-40). This was not a relaxation of holiness; it was its concentration. Love is not sentiment; it is obedience rooted in relationship. Christ did not abolish righteousness; He internalized it. What the law demanded externally, the Spirit now empowers internally.

This transition becomes clear at Pentecost. The Spirit descends not upon a single prophet but upon gathered believers (Acts 2:1-4). Fire rests on each one. The law once written on stone is now written on hearts (2 Corinthians 3:3). Ezekiel's prophecy of a new heart and a new spirit becomes lived reality (Ezekiel 36:26). Obedience shifts from compulsion to transformation; fear shifts to boldness; separation shifts to mission.

The church now operates under the law of the Spirit of life in Christ Jesus (Romans 8:2). This law frees from sin and death. Freedom does not mean moral looseness; it means moral empowerment. The Spirit produces fruit that the law could only command (Galatians 5:22-23). Love, joy, peace, patience, and self-control arise not from human willpower but from divine indwelling.

We must understand this movement from law to love carefully, especially in a technological culture obsessed with autonomy. Modern systems are designed to reduce friction, simplify decision-making, and increase personal control. That environment subtly conditions people to believe that transformation is mainly a matter of better systems, better habits, better tools, and better optimization. But the gospel exposes a deeper problem: humanity doesn't just need better management; it needs a changed heart.

Sin is not solved by efficiency; the Spirit conquers it. Tools promise self-sufficiency, systems promise optimization, and platforms promise access; yet none of these promises can cleanse desire. None of them can uproot pride, break lust, quiet envy, or awaken holy love for God. They may alter behavior for a season, but they cannot produce inward righteousness. The Spirit does not merely help people behave better; He transforms intention, reshapes appetite, and reorders affection. He teaches the heart to love what God loves and hate what God hates (Psalm 97:10).

That is why love is so central to the new covenant. Love fulfills what law could diagnose but not deliver. Love is not weakness; it is the evidence of the Spirit's rule in the inner life. It is holy, demanding, sacrificial, and obedient. In a culture increasingly mediated by screens and governed by convenience, Spirit-empowered love remains profoundly disruptive. It calls believers away from performance and back to communion. It demands presence in a distracted age, patience in an accelerated age, holiness in a permissive age, and dependence in a self-exalting age. The modern believer does not need stricter rules alone; he needs deeper surrender to the Spirit who fulfills righteousness from within. The church does not need a more polished law-centered image; it needs the reality of transformed hearts. Where the Spirit writes God's will within the believer, obedience stops feeling imposed and comes alive.

PENTECOST AND THE BIRTH OF A SPIRIT-DRIVEN MOVEMENT

Pentecost was not merely an emotional gathering; it marked the birth of the church as a Spirit-driven movement. The disciples were not strategizing branding or constructing outreach models; they were praying, waiting, unified, and obedient to Christ's instruction (Acts 1:12-14). When the Spirit came, the transformation was immediate and public.

Three thousand people responded to Peter's message (Acts 2:41). The same man who had denied Christ now proclaimed Him boldly. The difference wasn't improved rhetoric; it was empowerment. The Spirit gave utterance, conviction fell, repentance followed, baptisms multiplied, and community formed (Acts 2:4, 37-47). Teaching, fellowship, breaking of bread, and prayer became rhythms of life. Growth was not manufactured; it was organic and Spirit-initiated.

The pattern continued. Signs and wonders accompanied the proclamation; fear of the Lord produced reverence; generosity replaced greed; and boldness replaced hesitation (Acts 4:31-35). The church expanded across cultural and geographic lines. Persecution did not extinguish it; it scattered it, widening the gospel's reach (Acts 8:4). The Spirit turned opposition into expansion.

Pentecost established a principle that remains operative: power precedes effectiveness; prayer precedes impact; unity precedes multiplication. When the Spirit leads, growth occurs; when flesh leads, exhaustion follows. Modern churches may replicate structure, but cannot replicate Pentecost without surrender.

That distinction is urgent in the present age. Many churches now possess tools the early believers never imagined. Live broadcasts can reach millions; digital platforms can host discipleship globally; communication systems can organize ministries with stunning precision. Artificial intelligence can summarize information, accelerate workflows, and reduce administrative friction. None of those realities is evil in itself.

The danger arises when churches begin to mistake capability for power. Pentecost reminds the church that divine movement doesn't come from technical advantage. The church was not launched by superior infrastructure; the Spirit's descent launched it. It was not sustained by novelty; it was sustained by prayer, truth, boldness, holiness, and divine presence. This means the modern church must honestly evaluate itself. It is possible to have more reach than the early church and less fire. It is possible to have more content and less conviction, more systems and less surrender.

Without Pentecostal dependence, modern tools remain hollow channels. They may distribute a message, but they cannot breathe life into it. They may attract attention, but they cannot produce repentance. They may build an audience, but they cannot form the kind of community that the Spirit birthed in Acts. The Spirit is not intimidated by technology; He is grieved by self-reliance. When believers return to prayerful unity and obedience, modern tools become extensions of ancient power rather than substitutes for it. The church was born in fire, not formula. The Spirit, not strategy alone, drove its expansion. That same Spirit remains present. The modern church faces a simple question: will it rely on innovation as its strength, or will it return to the empowering presence that ignited its birth?

3

THE HOLY SPIRIT AND THE GROWTH OF THE CHURCH

Why True Expansion Cannot Be Engineered by Technology or AI

Church growth is one of the most discussed realities in modern ministry, yet it is also one of the most misunderstood. Many people treat growth as a numbers game: more seats filled, more giving, more volunteers, more social media engagement, and more campuses. Those things can indicate momentum, but they are not the definition of spiritual growth. A crowd can gather for entertainment, charisma, community, tradition, convenience, or habit without experiencing repentance, surrender, or transformation.

Scripture defines true growth differently. Real growth happens when people move from darkness to light, from sin to righteousness, and from spiritual death to spiritual life through Christ by the power of the Holy Spirit. This distinction matters because the church can appear successful while becoming spiritually hollow. It can expand its brand while shrinking its witness and increase its reach while decreasing its depth.

The New Testament portrait of growth is not primarily about size, but about salvation. When Acts records that believers were "added to the Lord," it highlights conversion rather than mere transfer. People were not merely joining an organization; they were being joined to Christ, with their allegiance shifting and their identity changing. As theologians like Simon Chan note, true Pentecostal growth is marked by this "ecclesial identity" formed by the Spirit. Their lives became evidence of the Spirit's intervention.

True growth also preserves loyalty to Christ's words. Jesus described Himself as the vine and His disciples as branches, making fruitfulness entirely dependent on abiding (John 15:5). That image eliminates the idea that ministry outcomes can be separated from spiritual union. A branch does not bear fruit by hustling; it bears fruit by remaining connected. The Spirit keeps that connection alive, nourishes the believer, convicts the heart, and forms Christ within the church. Disconnected growth may look energetic, but it will not remain spiritually fruitful.

This is where the modern church must exercise discernment. We live in an age where we can measure almost everything. Platforms track views, clicks, retention, reach, and engagement; ministries can monitor attendance patterns and donation habits with astonishing precision. While measurement and stewardship are useful, measurable activity and spiritual vitality are not the same thing. A church may know exactly how many people watched a sermon and still not know how many were pierced in conscience. It may count responses without discerning whether hearts were truly surrendered to Christ. Technology can report behavior, but only the Holy Spirit can transform the soul.

The danger is subtle because visible growth is easier to celebrate than invisible transformation. Numbers satisfy the flesh because they feel concrete and offer quick proof that something is happening. However, the Spirit often works deeper than what we can count. He breaks pride in private, calls people to repentance in secret, restores marriages behind closed doors, and purifies motives long before there is public evidence. Heaven values that kind of growth even when dashboards cannot record it.

For that reason, the church must refuse shallow definitions of expansion. Growth is not merely "bigger"; growth is holier, deeper, and more surrendered to Christ. Growth is seen when people begin to hate the sin they once defended and love the truth they once resisted. When the Holy Spirit is truly moving, He does not just increase attendance; He increases obedience.

ADDED TO THE LORD, NOT JUST ADDED TO THE ROLL

The language of Acts is precise and revealing. Scripture says, "the Lord added to the church daily such as should be saved" (Acts 2:47). That wording matters. The increase was God's work. The church proclaimed, prayed, and served, but the Lord produced the increase. This preserves humility in ministry; no leader, method, or system can claim ownership of spiritual harvest. The Spirit draws, convicts, regenerates, and incorporates believers into the body of Christ.

This is why church growth must never be reduced to mere transfer. If one congregation grows because discontented believers leave another, that may alter attendance patterns, but it is not the same as the kingdom advancing. The New Testament celebration was centered on conversion, discipleship, and the forming of a holy people, with new life standing at the center.

The early church was not built on spectacle; it was built on the Spirit's power. Peter preached, but God gave conviction; the apostles taught, but God gave illumination; the believers gathered, but God gave unity (Acts 2). Every part of the church's emergence pointed beyond human ability. That is why no honest reading of Acts allows the church to boast in methodology as though methodology were the source.

This speaks forcefully to the modern age. Churches can now build sophisticated outreach funnels, create polished media campaigns, and design high-level systems for guest engagement. These efforts may serve a purpose, and wise stewardship should not be mocked. But no follow-up sequence can regenerate a heart; no branding package can produce repentance; no communication strategy can make a dead soul alive. Only the Holy Spirit can do that.

A church may become excellent at assimilation while remaining weak in transformation. It may know how to retain attention without knowing how to lead people into surrender. It may learn to reduce friction for visitors without confronting the sin that separates people from God. This is where many ministries drift: they become skilled at gathering people without becoming equally burdened to see those people changed by the power of the Spirit. "Added to the roll" is administrative language; "added to the Lord" is redemptive language. One belongs to recordkeeping, the other to resurrection.

Growth that merely fills lists may impress people, but growth that joins souls to Christ glorifies God.

ABIDING, FRUITFULNESS, AND THE INNER WORK OF THE SPIRIT

Jesus did not describe kingdom fruitfulness in terms of frenzy; He described it as abiding (John 15:4). This image exposes a deep error in modern ministry: the assumption that constant activity guarantees spiritual productivity. The church can become addicted to motion—calendars fill, events multiply, and content increases—yet activity without abiding becomes noise. The Spirit does not reward busyness for its own sake; He produces fruit through communion with Christ.

Fruit is evidence of shared life. A church may produce polished content and strategic expansion, but if the life of Christ does not feed it through the Spirit, the fruit will eventually prove artificial. It may look persuasive for a season, but it will not endure. Real fruit carries the marks of heaven: love, holiness, truth, perseverance, humility, conviction, and worship (Galatians 5:22-23).

You cannot engineer, automate, or schedule the inner work of the Spirit like a campaign. The Spirit forms Christ in believers through conviction, surrender, discipline, and grace. He often works beneath the surface, uprooting hidden idols and healing wounds that distort obedience. This is why spiritual growth may look slower than institutional growth. Institutions can scale quickly, but souls do not heal on command. Systems can be duplicated overnight, but maturity cannot.

The technological age makes this tension sharper because digital culture trains people to expect immediate output. Delays feel like failure. But abiding is patient; it stays, listens, and submits. The Spirit teaches believers how to remain with God long enough for real fruit to emerge. Churches obsessed with speed often struggle with depth because the Spirit values root as much as fruit. The church must recover confidence in hidden growth: a man breaking with secret sin, a family learning to forgive, or a prideful leader repenting. These things do not always create headlines, but they reveal the life of Christ at work.

THE BOOK OF ACTS AND THE PATTERN OF SPIRIT-LED EXPANSION

Acts presents growth as the consequence of divine activity. The believers prayed, devoted themselves to the apostles' teaching, and shared fellowship (Acts 2:42). The Spirit moved with power, conviction fell, miracles occurred, and the Word spread. None of this happened because the apostles discovered a superior marketing model. It happened because heaven invaded earth through surrendered people.

The pattern is consistent: prayer precedes power, power precedes witness, and witness precedes increase. The church began with obedience and waiting on God, not expansion plans. When the Spirit came, sustained transformation followed. Peter's preaching was empowered, not merely eloquent.

Acts also shows that true growth can survive hostility; persecution scattered the church, which only widened the reach of the gospel (Acts 8:1, 4). That resilience reveals the difference between Spirit-born growth and flesh-built momentum. Flesh-built momentum often collapses under pressure, while Spirit-born growth deepens. The deepest barrier to growth is not inconvenience, but sin; the deepest need is not better presentation, but regeneration; and the deepest source of momentum is the Holy Spirit.

Strategy and planning have their place (Acts 6:1-7), but they must remain subordinate to divine direction. In a ministry following the pattern of Acts, we will care about doctrine, prayer, holiness, and witness. It will not confuse structure with life, understanding that true expansion begins when the Spirit breathes on the ordinary work of faithful believers.

MEASURABLE GROWTH AND THE ILLUSION OF SUCCESS

One of the most dangerous temptations facing the modern church is the illusion that what can be measured is what matters most. Metrics can reveal patterns and support stewardship, but problems arise when metrics become masters rather than tools. We can count views, but we cannot count repentance; we can chart attendance, but we cannot graph purity of heart.

When ministries become overly fascinated with measurable success, they may begin shaping everything around what performs best rather than what forms people best. Sermons may grow shorter to avoid losing attention, and messaging may become "safer" and less prophetic because offense threatens retention. Leaders may interpret growth curves as divine confirmation even when prayer is thinning, and holiness is weakening.

The church must refuse to let performance culture define success. A ministry can be highly visible and deeply compromised. The Spirit values integrity over image and produces fruit, not facade. Where image management dominates, anxiety increases. Where the Spirit governs, peace and boldness stabilize the work.

Discernment is essential: not every open door comes from God, and not every viral moment proves divine favor. The Spirit may prune before He multiplies and expose what is unhealthy before He expands what is holy. Measured success has its place, but it cannot occupy the throne. The church must ask deeper questions: Did people encounter truth? Were hearts convicted? Did holiness deepen? These questions move the ministry back into spiritual reality.

WHY TRUE EXPANSION CANNOT BE ENGINEERED

True expansion cannot be engineered because the church is not a machine; it is a living body. As Paul writes, "I have planted, Apollos watered; but God gave the increase" (1 Corinthians 3:6). That statement dismantles human pride and guards against despair. Engineering can produce systems and processes, but it cannot create brokenness over sin or hunger for righteousness. It cannot reveal Christ to the heart with saving power or breathe life into dry bones. Those realities belong to the Holy Spirit's ministry alone.

Artificial intelligence sharpens this distinction. AI can imitate tone, summarize doctrine, and accelerate production. It can assist pastors and mission teams in practical ways, reducing friction. Yet it cannot produce revelation, conviction, or holy transformation. It can generate language that sounds spiritual, but it cannot generate the "new birth" (John 3:3-8).

The more persuasive a tool becomes, the easier it is to confuse assistance with authority. AI can sound spiritual yet lack reverence; it can offer instant output that tempts leaders to bypass prayerful dependence.

Speed is not wisdom, and convenience is not guidance. When the church treats instant output as a substitute for discernment, it exchanges intimacy for immediacy. Tools must remain tools; the Spirit must remain Lord over process, motive, and method.

GROWTH THAT GLORIFIES GOD

Growth that glorifies God keeps Christ at the center, the Spirit as the source, and holiness as the evidence. It does not despise organization, but it does not worship it. It welcomes tools without kneeling to them.

A Spirit-grown church will preach with conviction, love that costs something, and holiness that survives private testing. It will form disciples, not just consumers, and value prayer as a necessity.

This kind of church reminds the world that life is still sacred and truth is still personal. It shows that optimization alone does not solve humanity's deepest needs. Souls do not merely need better systems; they need the Spirit of the living God.

The church should thank God for every genuine increase and plan wisely, but it must remember that none of these can substitute for divine breath. Real growth still comes the old way: through prayer, truth, surrender, and the living power of the Holy Spirit.

4

THE SPIRIT IN THE OLD TESTAMENT: FOUNDATION FOR TODAY

Ancient Power Shaping Faithfulness in a Digital Civilization

The first dimension of the Holy Spirit revealed in the Old Testament is cosmic. Before Israel exists, before the covenant is formalized, and before prophecy is spoken, the Spirit is active in creation. Genesis records that "the earth was without form, and void; and darkness was upon the face of the deep. And the Spirit of God moved upon the face of the waters" (Genesis 1:2). That movement is not decorative language; it signals divine involvement in ordering chaos. The Spirit does not appear after structure is established; He appears before it, present when there is nothing but disorder.

.

This cosmic dimension establishes a foundational truth: the Spirit is not limited to religious settings. He governs creation itself. Job affirms that "The Spirit of God hath made me, and the breath of the Almighty hath given me life" (Job 33:4). The Psalms declare that when God sends forth His Spirit, creation occurs and the face of the earth is renewed (Psalm 104:30). Creation is not a closed system running independently of God; His Spirit sustains it. Life is not self-generating; it depends on the Spirit.

The Spirit's cosmic role also includes providence. Scripture portrays Him as directing natural forces, restraining chaos, and sustaining order. When Psalm 139 asks, "Whither shall I go from thy spirit? or whither shall I flee from thy presence?" (Psalm 139:7), it affirms omnipresence. The Spirit is not geographically confined; He is active in every realm of existence.

Recognizing the Spirit's cosmic role keeps believers from reducing Him to church activity alone. The Spirit is present in the field, the storm, the city, the wilderness, the boardroom, and every environment where life continues by divine permission. That truth carries weight in a digital civilization where many experience reality through layers of screens and metrics. The Old Testament refuses a split between the mechanical and the spiritual; the Spirit is active in material reality itself. He is not detached from history, weather, breath, or time.

The cosmical dimension also builds confidence. If the Spirit orders galaxies and sustains life, He is not threatened by digital networks or artificial intelligence. The same Spirit who hovered over primordial waters oversees modern innovation. Nothing emerges outside His awareness; nothing progresses beyond His authority. The Spirit does not react to history; He directs it toward redemptive ends.

THE THEOCRATIC DIMENSION: GOD GOVERNING HIS PEOPLE

The second dimension revealed in the Old Testament is the theocratic dimension. Here, the Spirit empowers individuals to lead, govern, and fulfill covenantal responsibility. The Spirit comes upon judges such as Othniel, Gideon, and Jephthah to deliver Israel from oppression. These leaders are not self-appointed revolutionaries; they are Spirit-empowered instruments within God's redemptive plan. Their authority flows from divine enablement, not from personal ambition.

The Spirit also empowers craftsmanship and administration. Bezalel was filled with the Spirit of God "in wisdom, and in understanding, and in knowledge, and in all manner of workmanship" to construct the tabernacle (Exodus 31:3). This detail is significant: the Spirit not only empowers prophets and warriors; He also empowers artisans and organizers. Creativity and craftsmanship become expressions of the Spirit's activity. He is present where beauty is shaped, where sacred order is built, and where practical skill serves divine purpose.

Kingship also falls under this dimension. When Samuel anoints David, "the Spirit of the Lord came upon David from that day forward" (1 Samuel 16:13). Leadership in Israel is spiritual; the Spirit legitimizes authority. When Saul disobeys, the Spirit departs. This shows that divine empowerment is not a permanent possession but depends on obedience. The Spirit is not manipulated by position; He responds to surrender.

Prophetic ministry flourishes within the theocratic dimension. Holy men of God spoke as they were "moved by the Holy Ghost" (2 Peter 1:21). The authority of prophecy rests in the Spirit's initiative. This dimension establishes a principle for the modern church: leadership without Spirit empowerment becomes bureaucratic. Skill without Spirit guidance becomes self-exalting. In an age saturated with management systems, the church must remember that genuine authority flows from the Spirit's commissioning. Modern churches can become administratively refined while quietly losing the spiritual authority that makes ministry alive. Ancient Israel's story warns against mistaking office for anointing; a title did not guarantee God's backing.

THE INDIVIDUAL DIMENSION: GOD TRANSFORMING THE HEART

The third dimension in the Old Testament reveals the Spirit working within the individual. David's plea in the Psalms is deeply personal: "Create in me a clean heart, O God; and renew a right spirit within me. Cast me not away from thy presence; and take not thy holy spirit from me" (Psalm 51:10-11). Moral reform cannot be self-generated; God must impart it.

Ezekiel intensifies this promise, declaring that God will "take away the stony heart out of your flesh, and I will give you a heart of flesh" (Ezekiel 36:26). God declares He will put a new spirit within His people so that they walk in His statutes. This shifts the focus from external compliance to internal renewal. The Spirit does not merely inform behavior; He reshapes desire, alters motivation, and transforms the will from within.

Joel extends the promise further, declaring that "I will pour out my spirit upon all flesh; and your sons and your daughters shall prophesy, your old men shall dream dreams, your young men shall see visions" (Joel 2:28). The Spirit's work would no longer be confined to select leaders; it would become democratized. This anticipates an era where the Spirit dwells within believers.

The Spirit also functions as teacher and guide. Nehemiah recounts that God gave His "good spirit to instruct them" (Nehemiah 9:20). Wisdom in Scripture depends on the Spirit; insight comes through spiritual illumination. This remains essential because, in a world saturated with information, the deepest crisis is a hard heart. Data can inform behavior, but it cannot regenerate desire; only the Spirit performs inward renovation.

The Spirit still addresses motive and addresses hidden rebellion, pride, and self-deception. The church must not allow technological polish to cover spiritual poverty. Transformation is not a matter of trying harder; the heart of stone does not soften by self-effort alone. The soul is not healed by information; God must breathe on it.

PROGRESSIVE REVELATION AND THE ROAD TO PENTECOST

The Old Testament lays the foundation for the doctrine of the Spirit, which unfolds with increasing clarity. Isaiah prophesies about a coming Messiah upon whom "the spirit of the Lord shall rest... the spirit of wisdom and understanding, the spirit of counsel and might, the spirit of knowledge and of the fear of the Lord" (Isaiah 11:2). These descriptions anticipate Jesus, revealing that the Spirit's gifts are purposeful and Christ-centered.

When Jesus arrives, He embodies this prophecy. He announces that "The Spirit of the Lord is upon me, because he hath anointed me to preach the gospel to the poor" (Luke 4:18). The Old Testament expectation converges in Christ.

This convergence prepares the way for Pentecost, where the Spirit's work moves from selective empowerment to widespread indwelling.

Understanding this progression guards against historical amnesia. The Spirit has been active from the beginning—creating, empowering, convicting, and guiding. The technological age does not introduce a new Spirit; it continues under the same Spirit. What is recent is often assumed to be more advanced, but Scripture reverses that logic. A deeper understanding of ancient foundations secures the church's future. The Spirit is the one constant reality in a changing world. He orders chaos, empowers leadership, transforms individuals, and advances redemptive history. Tools evolve, and empires fall, but the Spirit remains.

5

THE HOLY SPIRIT IN THE NEW TESTAMENT CHURCH

Indwelling Presence and Living Authority Beyond Algorithms

As Jesus neared the end of His ministry, He did not give His disciples strategy manuals. He left them with a promise. He said the Father would send another Comforter. The Greek word is *Parakletos*. It means one called alongside to help, advocate, counsel, or intercede. The promise was not optional encouragement; it was an essential provision. Jesus told them plainly that it was better for Him to go away so that the Spirit could come. He stated, "It is expedient for you that I go away: for if I go not away, the Comforter will not come unto you; but if I depart, I will send him unto you" (**John 16:7**). That statement reveals the magnitude of the Spirit's role in the life of the church.

The disciples did not initially understand the weight of this promise. They had walked with Jesus physically. They had seen Him teach, heal, confront demons, and calm storms. Physical proximity felt like security. Yet Jesus declared that the Spirit's coming would be greater for them than His continued bodily presence. This shift signaled a transition from external leader-
ship to internal empowerment. The Spirit would not

merely visit them; He would dwell within them. Christ's ministry would no longer be confined to one region, one road, one house, or one gathered crowd. Through the Spirit, His presence would become immediate and personal within believers themselves.

The promise also revealed continuity. The Spirit would not introduce a new message disconnected from Christ. Jesus said the Spirit would bring to remembrance what He had spoken (**John 14:26**). The Spirit would not speak independently or contradict the Son. As noted by theologians like **Gordon Fee**, the Spirit is the "personal presence of God," and His primary role is to glorify Christ and reveal what belongs to the Father (**John 16:14**). This unity prevents spiritual confusion. Any movement that claims to be Spirit-led yet diminishes Christ stands outside this promise. The Holy Spirit does not compete with Jesus for attention. He magnifies Him, interprets Him rightly, and presses His truth into the believer's conscience.

The promise of the Comforter also addresses fear. Jesus acknowledged that His followers would face hostility. They would be misunderstood, persecuted, and opposed. The Spirit would not eliminate hardship; He would strengthen them within it. The Comforter would teach, guide, and sustain. The disciples would not be abandoned; they would not be left as orphans (**John 14:18**). The Spirit's presence would make Christ's mission portable. It would travel through believers into every environment, every city, every generation, and every form of cultural change.

That promise still burns with urgency in the modern church. Believers live in a world saturated with noise, pressure, reaction, and distraction. Public opinion can shift in hours. Digital hostility can form in minutes. Truth can be challenged, misrepresented, or mocked on a scale. Yet the Spirit remains God's abiding presence with His people. He is still the Advocate in hostile spaces, the Teacher in moments of confusion, and the Comforter in seasons of grief. The church does not face a connected world alone; it carries God's indwelling presence within it.

This is where the modern church must be careful. It is easy to admire the Spirit conceptually while relying on other things functionally. Churches can build systems, staff teams, content pipelines, media strategies, and digital infrastructures, yet quietly treat the Spirit as a theological ornament rather than the living source of guidance and power. Jesus did not promise His disciples a better method; He promised them His Spirit. That order must remain intact. The church may use many tools, but it must trust one source.

THE SPIRIT OF TRUTH AND GUIDANCE

Jesus described the Holy Spirit as the Spirit of truth (**John 16:13**). This designation is not abstract. Truth in Scripture is relational and redemptive. It is not merely factual accuracy; it is alignment with God's revealed will. The Spirit guides believers into all truth. He does not overwhelm them with information they cannot bear. He reveals progressively, in harmony with divine wisdom. Guidance in the Spirit is not random inspiration; it is holy direction rooted in the character and purposes of God.

The Spirit's guidance includes conviction of sin, clarification of righteousness, and assurance of judgment (**John 16:8**). Conviction is not condemnation; it is correction aimed at restoration. The Spirit exposes unbelief and reveals Christ as the remedy. He clarifies what righteousness looks like in light of Christ's ascension. He assures believers that evil will not have the final word because the prince of this world stands judged (**John 16:11**). This threefold ministry anchors the church in moral clarity amid cultural confusion. Without this guidance, people drift into self-defined spirituality and call it freedom.

Guidance also operates in practical decisions. The Acts records multiple instances in which the Spirit directed missionary movements. He forbade entry into certain regions and redirected the apostles elsewhere (**Acts 16:6-7**). This detail shows that guidance is not always expansion in the direction of human preference. Sometimes the Spirit restrains. Sometimes He redirects. Sometimes He closes what appears open; sometimes He leads toward what appears difficult. Obedience requires attentiveness, humility, and the willingness to let divine wisdom overrule human preference.

The Spirit of truth also guards against deception. In a

technological culture saturated with information, distortion spreads quickly. False teaching can circulate globally within minutes. Emotional narratives can replace biblical doctrine. Confident voices can sound persuasive while being spiritually empty. The Spirit counters this through illumination. He opens Scripture. He brings remembrance of Christ's words. He exposes the error. He anchors believers in revealed truth. Without this ministry, the church becomes vulnerable to every cultural wind and every polished deception dressed in spiritual language.

This guiding presence requires surrender. The Spirit does not coerce obedience; He leads. Those who are led by the Spirit are called sons and daughters of God (**Romans 8:14**). Leadership implies relationship. Guidance implies responsiveness. When believers ignore conviction or resist correction, clarity fades. When they yield, understanding deepens. The Spirit of truth does not merely inform the mind; He aligns the heart with God's purposes. He trains believers not just to know what is true, but to love it enough to obey it.

This is where the contrast with algorithmic culture becomes sharp. Algorithms sort preferences. They predict habits. They amplify patterns. They can recommend content based on prior behavior, but they cannot lead the soul into holiness. They cannot confront hidden sin with divine tenderness. They cannot distinguish between what is appealing and what is righteous. The Spirit does not merely respond to human preference. He challenges it, purifies it, and redirects it. In an age when recommendation engines and digital reinforcement loops shape many people, the church must remember that the Spirit's guidance is not reactive. It is authoritative. He leads according to heaven's wisdom, not human appetite.

The Spirit also guides in ways that often conflict with the speed of modern life. Digital systems reward immediacy. The Spirit often leads progressively. Platforms reward hot takes and instant responses. The Spirit often cultivates restraint, silence, patience, and depth. What quickly trends is not always what transforms deeply. The church must learn again how to wait, discern, listen, test, and pray. The Spirit of truth is not in a hurry to satisfy the flesh. He is committed to forming Christ in His people.

THE SPIRIT AND NEW BIRTH

Jesus' conversation with Nicodemus reveals the heart of the Spirit's regenerative work. "Except a man be born of water and of the Spirit, he cannot enter into the kingdom of God" (**John 3:5**). New birth is not a metaphorical exaggeration; it is a spiritual reality. Flesh gives birth to flesh; Spirit gives birth to spirit (**John 3:6**). This statement establishes a boundary. Religious knowledge alone cannot produce new life. Moral reform alone cannot produce new life. Institutional belonging alone cannot produce new life. The Spirit alone regenerates.

New birth transforms identity. When the Spirit indwells a believer, adoption occurs. The Spirit bears witness with our spirit that we are children of God (**Romans 8:16**). This witness is internal assurance, not external applause. The Spirit does not merely modify behavior; He confirms belonging. Believers cry, "Abba, Father," not as repetition but as relationship (**Galatians 4:6**). This filial intimacy was not available under the old covenant in the same way. The Spirit makes it experiential. He turns theology into life and doctrine into communion.

Regeneration also breaks sin's dominion. Paul writes that the law of the Spirit of life in Christ Jesus sets believers free from the law of sin and death (**Romans 8:2**). This freedom is not license; it is liberation from mastery. The Spirit empowers believers to put to death the deeds of the body (**Romans 8:13**). Sanctification is not self-discipline alone; it is cooperation with the indwelling Spirit. The new birth is not a spiritual badge. It begins an entirely new life order, in which the Spirit confronts old desires and forms new ones.

The fruit of the Spirit reveals the nature of this new life. Love, joy, peace, patience, kindness, goodness, faithfulness, gentleness, and self-control emerge as evidence (**Galatians 5:22-23**). These qualities are not personality traits; they are Spirit-produced virtues. In a culture that values visibility and speed, fruit grows quietly and consistently. The Spirit cultivates character beneath the surface. He works where cameras don't reach and where public validation doesn't matter. That hidden work is often the most convincing proof that the Spirit is present.

The necessity of new birth remains urgent. Digital access

to sermons does not equal regeneration. Church attendance does not equal transformation. Familiarity with Christian language does not equal indwelling presence. A person may know how to quote Scripture, navigate spiritual spaces, and speak in approved religious phrases while remaining spiritually unchanged. The Spirit's regenerative work remains the dividing line between religious familiarity and spiritual life. Without it, the church becomes an organization of informed participants. With it, the church becomes a living body animated by divine breath.

This is one of the most serious challenges facing the church in a connected age. Technology can distribute religious content endlessly. It can create the impression of spiritual saturation while leaving the heart untouched. A person can listen constantly and remain unconverted. A church can broadcast beautifully and still lack evidence of new birth among its people. Access cannot replace the Spirit's work. No amount of digital proximity to truth can substitute for inward regeneration. Information may prepare the way; only the Spirit raises the dead.

New birth also restores moral seriousness. The Spirit does not merely make people feel spiritual. He creates holy dissatisfaction with sin. He awakens hunger for righteousness. He breaks the numbness that allows people to live divided lives without grief. In a world increasingly comfortable with curated identities and fragmented selves, the Spirit creates integrity. He brings the hidden life into the light. He makes holiness desirable. He teaches the believer to hate what destroys and to love what pleases God. That is the mark of real transformation.

ROMANS 8 AS THE BLUEPRINT FOR SPIRIT-LED LIVING

Romans 8 stands as one of the clearest summaries of the Spirit's ongoing ministry. It opens with a declaration of no condemnation for those in Christ Jesus (**Romans 8:1**). This freedom rests not on human perfection but on the Spirit's application of Christ's work. The law was weakened by flesh. God accomplished what the law could not by sending His Son and condemning sin in the flesh (**Romans 8:3**). The Spirit now empowers believers to walk not according to the flesh but according to Himself (**Romans 8:4**). This chapter does not offer theory; it gives the church a living framework for daily life under the rule of the Spirit.

The chapter contrasts two mindsets. The flesh mindset results in death (Romans 8:6). The Spirit mindset results in life and peace. This contrast reveals daily dependence. Spirit-led living is not a one-time decision; it is a continual orientation. The Spirit shapes desires, thoughts, and responses. When believers yield, peace follows; when they revert to fleshly control, conflict intensifies. Romans 8 forces a choice between self-governance and Spirit-governance. The issue is not whether a person has religious interest; the issue is what governs the inner life.

Romans 8 also addresses suffering. Present sufferings are not worthy to be compared with future glory (**Romans 8:18**). Creation itself groans, awaiting liberation. Believers groan inwardly, awaiting redemption (**Romans**

8:23). In this tension, the Spirit intercedes with groanings too deep for words (**Romans 8:26**). This intercession is intimate and active. The Spirit helps in weakness. He does not abandon believers when they are confused. He prays on their behalf according to God's will. When language fails, the Spirit does not. When believers are exhausted, disoriented, or burdened beyond expression, the Spirit remains active in their weakness.

The chapter climaxes with assurance. If God is for us, who can be against us? (**Romans 8:31**). Nothing can separate believers from God's love in Christ. Neither death nor life, angels nor powers, present nor future (**Romans 8:38-39**). This assurance is not motivational rhetoric; it is covenantal security grounded in the Spirit's presence. The Spirit seals believers unto redemption (**Ephesians 4:30**). He guarantees inheritance. He steadies the soul against fear and against the unstable conditions of the present age.

Romans 8 functions as a blueprint because it integrates freedom, identity, suffering, intercession, and victory under the Spirit's governance. It shows that Spirit-led living is not confined to ecstatic experience. It encompasses daily obedience, patient endurance, confident hope, and unwavering assurance. In a technological age filled with instability, comparison, distraction, and emotional overload, this blueprint provides stability.

The Spirit anchors believers in truth, sustains them through suffering, and assures them of ultimate victory.

It also answers a subtle modern temptation. Many people want guidance without submission, peace without holiness, confidence without repentance, and spiritual language without spiritual surrender. Romans 8 will not permit that shallow arrangement. Life and peace belong to the Spirit-governed mind. Assurance belongs within a life shaped by the Spirit's witness and Christ's work. There is no shortcut around surrender. The flesh cannot be optimized into holiness; it must be crucified (**Galatians 2:20**). The Spirit does not merely improve the old life; He governs a new one.

This matters for the church because Romans 8 describes more than private devotion. It describes the kind of people the church must become to remain alive in a distracted civilization. A church shaped by Romans 8 will not simply talk about freedom; it will walk in it. It will not merely celebrate victory in language; it will endure suffering with hope. It will not confuse outward momentum with inward life. It will know that the deepest stability in unstable times is not institutional strength, digital reach, or polished communication. It is the indwelling Holy Spirit.

The New Testament church thrived not because it possessed infrastructure, but because it possessed indwelling power. The promise of the Comforter, the guidance of the Spirit of truth, the miracle of new birth, and the blueprint of Romans 8 remain operative. The Spirit continues to apply Christ's work, guide believers, regenerate hearts, and sustain hope. Technology may alter communication, but it does not alter this foundation. The church's vitality still rests on the same Spirit who descended at Pentecost and who continues to dwell within believers today.

6

PENTECOSTAL REVIVAL AND THE MODERN MOVEMENT OF THE SPIRIT

Spiritual Outpouring, Global Awakening, and the Limits of Automation

The Day of Pentecost stands as one of the most decisive moments in Christian history. Fifty days after Christ's resurrection, the disciples gathered together in one place.

They were not organizing campaigns. They were not negotiating influence. They were waiting in obedience to Jesus' instruction. Then suddenly there came a sound from heaven as of a rushing mighty wind, and it filled the house where they were sitting. Cloven tongues like fire appeared and rested upon each of them. They were all filled with the Holy Ghost and began to speak with other tongues as the Spirit gave them utterance.

This was not emotional excess; it was fulfillment. Joel had prophesied that God would pour out His Spirit upon all flesh (Joel 2:28). Sons and daughters would prophesy; young men would see visions; old men would dream dreams. The Spirit would no longer be limited to selective visitation; He would be poured out. Pentecost inaugurated that promise.

What had been anticipated in the prophets became

realized in the church. The immediate result was a proclamation. Peter stood and preached Christ crucified and risen. The same man who had denied Jesus now spoke with boldness and clarity. His transformation cannot be explained by improved confidence; it must be attributed to empowerment by the Spirit. The Spirit not only filled them; He emboldened them. He gave utterance. He made Christ central.

Three thousand souls were added that day (Acts 2:41). The church was born not through institutional design but through divine intervention. Pentecost also established a pattern. The Spirit-filled church devoted itself to the apostles' teaching, fellowship, breaking of bread, and prayer (Acts 2:42). Signs and wonders accompanied their witness. Awe came upon every soul. Generosity flowed naturally; community formed organically; growth occurred daily. This was not a single-day anomaly; it was the beginning of a Spirit-sustained movement.

Pentecost's continuing impact remains evident. The Spirit still fills. The Spirit still empowers. The Spirit still gives utterance. The Spirit still forms community. Outward expressions may differ across cultures and centuries, but the core reality remains unchanged. Pentecost was not a historical curiosity; it was the divine release of power for the church's witness in every age. It remains the clearest reminder that spiritual life begins not with strategy, but with surrender.

That truth speaks forcefully to the present moment. The

modern church has tools the apostles never imagined. Messages can be distributed globally in seconds. Worship can be streamed across continents. Teaching can be archived, clipped, translated, and shared at scale. Yet none of those developments alter the basic order established at Pentecost. Power still comes before effectiveness. Prayer still comes before impact. Divine presence still comes before real expansion. A church can possess advanced tools and still lack Pentecostal fire. It can be visible without being vital. It can be efficient without being alive. Pentecost remains the rebuke of every attempt to replace divine empowerment with human capability.

THE RISE OF PENTECOSTALISM AND CHARISMATIC RENEWAL

At the dawn of the twentieth century, a renewed emphasis on Spirit baptism and spiritual gifts began to spread. The Holiness movement in the United States laid the groundwork for a deeper hunger for sanctification and spiritual experience. Charles Fox Parham taught that speaking in other tongues was a sign of baptism in the Holy Spirit. In 1906, the Azusa Street revival in Los Angeles became a catalytic event. Worship was interracial and spontaneous, marked by intense prayer and spiritual manifestations.

This revival did not remain localized. It spread rapidly across denominational lines and geographic boundaries. Churches formed that emphasized the gifts of the Spirit, divine healing, prophecy, and expressive worship. Pentecostal denominations such as the Church of God in

Christ, the Assemblies of God, and others emerged. Their growth was not confined to North America; it spread rapidly into Latin America, Africa, and Asia.

The Charismatic movement of the 1960s further broadened this renewal. Spiritual gifts began to be practiced within mainline Protestant and Roman Catholic congregations. Dennis Bennett's public testimony of speaking in tongues became a turning point for many churches. The Spirit's activity was no longer perceived as restricted to one stream of Christianity. It began crossing institutional boundaries.

This historical rise matters for reasons beyond description. It shows that the hunger of Acts did not disappear after the apostolic era. The modern church did not invent longing for the Spirit; it rediscovered it. Pentecostalism and charismatic renewal reopened questions that many formal traditions had left dormant. Can God still heal? Does the Spirit still empower with gifts? Is supernatural ministry still part of normal Christian expectation? These movements answered yes, not merely in doctrine, but in lived practice.

Their spread also revealed something about spiritual appetite. People were not only seeking better systems, sharper arguments, or stronger institutions. Many were seeking an encounter. They wanted evidence that God was present, active, and near. They wanted worship that breathed, prayer that expected, and preaching that carried fire rather than form alone. Whether every expression was mature is another question; renewal movements have always required discernment. But their very rise testified that large segments of the church were unwilling to settle for a version of Christianity that had order without power.

This is one reason the chapter matters so much in the setting of technological modernity. The rise of Pentecostal and charismatic renewal exposed a tension that still exists now. Churches can become highly structured, intellectually informed, and institutionally stable, while people inside them remain hungry for a living encounter with God. The answer to that hunger is not chaos, anti-intellectualism, or spectacle. It is renewed surrender to the Spirit who makes Christ real and ministry alive.

THE ONGOING MINISTRY OF THE HOLY SPIRIT IN REVIVAL MOVEMENTS

Pentecostalism emphasized immediacy. It affirmed that the Spirit's power was not reserved for the first century. It stressed personal encounters, experiential worship, and expectancy of supernatural intervention. Critics questioned excesses and emotionalism; supporters testified to transformed lives, healed bodies, and renewed faith. Regardless of perspective, the movement undeniably reshaped global Christianity.

The significance of this rise lies not merely in the numbers but in the theological emphasis. Pentecostal revival re-centered the Holy Spirit in practical ministry. It insisted that the Spirit was not a doctrine to be affirmed but a presence to be experienced. It challenged churches that had grown formal and reminded them that Acts was not closed. The Spirit who descended at Pentecost had not retired.

Healthy revival movements do more than excite emotion. They restore proportion. They force the church to confront what it has begun to trust. They expose the difference between ceremonial religion and living power. They remind believers that spiritual gifts are not museum pieces, that prayer is not filler, and that holiness is not optional ornamentation for ministry. The Spirit's ongoing ministry in revival has always aimed to renew reverence, intensify witness, and call the church back to dependence.

This is why revival cannot be reduced to atmosphere alone. True revival produces repentance, renewed love for Scripture, deeper prayer, moral seriousness, communal transformation, and an intensified witness to Christ. It does not merely stir a room; it alters a people. It restores hunger where routine has taken over. It breaks pride where self-sufficiency has settled in. It turns the church away from admiration of its own machinery and back toward the God who gives life.

Revival history also warns the church not to confuse manifestation with maturity. Fire without formation can become instability. Zeal without truth can drift into error. Claims of power without godly character can produce damage rather than renewal. That is why the Spirit's ongoing ministry must always be grounded in biblical order, doctrinal clarity, and Christ-centered humility. The answer to misuse is not rejection of the Spirit's work; it is better discernment, deeper shepherding, and greater submission to Scripture.

The modern church needs this reminder because technological culture creates its own kind of counterfeit revival. Viral attention can mimic momentum. Emotional content can imitate spiritual intensity. Large-scale reactions can look like awakening. Yet revival born of the Spirit carries fruit that no platform can manufacture. It convicts more than it performs. It sanctifies more than it entertains. It glorifies Christ more than it magnifies personalities. Where the Spirit truly moves, the center of gravity shifts away from spectacle and back to surrender.

GLOBAL GROWTH IN AFRICA, ASIA, AND THE GLOBAL SOUTH

The dramatic expansion of Christianity in the Global South is one of the most significant developments of the modern era. Africa, in particular, witnessed exponential growth throughout the twentieth century. From a small Christian population at the start of the century, the continent saw tens of millions profess faith in Christ. Reports indicate thousands being added daily in various regions.

This growth cannot be attributed solely to missionary strategy. Cultural dynamics, social upheaval, and economic transitions played roles. Yet the unifying thread in many African movements was a strong emphasis on the Holy Spirit. Prayer gatherings flourished. Worship was expressive. Deliverance ministry was common. They did not treat the Spirit as an abstract theological concept but as an active presence confronting evil and healing brokenness.

In Asia, particularly in parts of China and South Korea, underground churches multiplied despite governmental resistance. Prayer movements grew; house churches expanded; testimonies of healing and spiritual encounter became catalysts for conversion. The Spirit's work did not depend on public buildings or institutional security; it thrived in adversity. Latin America experienced similar revival currents. Pentecostal and Charismatic churches expanded rapidly. Communities previously shaped by nominal Christianity encountered a renewed emphasis on personal conversion and Spirit empowerment. Worship styles adapted to local culture while retaining focus on the Spirit's immediacy.

The demographic shift of Christianity's center from the Northern Hemisphere to the Southern Hemisphere underscores an important principle. The Spirit is not bound to one culture; He moves where hunger exists. He empowers communities that respond in prayer and surrender. This global movement continues the pattern seen in the book of Acts. The Spirit continues to expand the church across languages, ethnicities, and social classes.

The significance of this global expansion is sharpened by the age in which it is happening. A connected world can now witness revival currents in real time. Testimonies move across borders quickly. Prayer burdens spread internationally. Technology can help the church see what the Spirit is doing in other regions, but it cannot cause those awakenings. The global South is not growing because it discovered better branding. In many places, it is growing because prayer has remained central, spiritual hunger has remained visible, and dependence upon the Spirit has remained nonnegotiable.

This should confront assumptions in more affluent contexts. Wealth, infrastructure, and polished systems do not guarantee vitality. In fact, they can sometimes conceal spiritual dullness. Meanwhile, communities with fewer resources often display greater expectation, deeper corporate prayer, and stronger awareness of spiritual realities. The lesson is not that poverty is holy; the lesson is that dependence matters. The Spirit often finds clearer room to move where self-sufficiency has been exposed as fragile.

WHY THE SPIRIT THRIVES WHERE HUNGER IS HIGH

Spiritual hunger often correlates with spiritual growth. Regions experiencing poverty, political instability, or social fragmentation frequently display openness to the gospel. This does not romanticize suffering; it recognizes that desperation often produces receptivity. When material security is limited, spiritual need becomes visible.

The Spirit moves powerfully where people recognize their dependence.

Pentecostal and Charismatic expressions often flourish among marginalized communities. The emphasis on healing speaks to physical need; the emphasis on deliverance speaks to spiritual oppression; the emphasis on communal worship addresses isolation. The Spirit meets people where they live. He does not require cultural uniformity; He engages contextual struggles with divine authority.

This does not mean that prosperity inhibits revival; it means comfort can dull urgency. In affluent contexts, distraction competes with devotion. Technology fills silence; entertainment occupies attention. The Spirit does not withdraw in such settings, but He is often resisted by self-sufficiency. Where hunger increases, responsiveness follows.

The Spirit thrives where prayer is central. Revival movements consistently reveal intense prayer practices. Corporate gatherings, all-night intercession, fasting, and repentance mark seasons of renewal. Prayer signals dependence: dependence invites empowerment; empowerment produces witness; witness produces growth. The cycle is not mechanical; it is relational.

The lesson for the modern church is clear: cultivate hunger. Programs do not substitute for prayer. Production quality does not substitute for presence.

Strategic excellence does not substitute for surrender. The Spirit moves where Christ is exalted, where repentance is genuine, and where prayer is persistent.
This is where automation's challenge becomes deeply spiritual. Automation promises a frictionless process. It reduces waiting, compresses labor, and rewards speed. But hunger is rarely formed through convenience. Hunger is formed in prayer, in need, in longing, in holy dissatisfaction, and in sustained exposure to God's presence. The Spirit rarely builds deep dependence through shortcuts; He builds it through yieldedness. A church trained only by convenience may become efficient without becoming desperate for God. A church that still knows how to tarry, fast, repent, and cry out will remain open to the kinds of visitations that cannot be engineered.

The modern church must therefore ask itself hard questions. Are we seeking God, or merely improving process? Are we building prayer altars, or only better workflows? Are we cultivating reverence, or only refining presentation? Hunger is not sentimental language; it is a spiritual condition. It is the refusal to be satisfied with ministry that looks alive but lacks divine breath. It is recognizing that no amount of technological sophistication can replace a fresh move of the Spirit. Pentecostal revival and global expansion show that the Spirit's mission continues vigorously today. The same Spirit who empowered the early church now animates believers across continents. He crosses linguistic barriers; He transcends cultural divisions; He operates beyond institutional limitations. In a technological age, this global movement becomes even more interconnected, yet the source remains unchanged. Growth that endures is growth born of the Spirit.

7

TECHNOLOGY, MEDIA, AND THE MISSION OF GOD

Using Digital Tools Without Replacing Spiritual Dependence

The church has never carried out its mission without tools. From the beginning, the people of God have utilized the available means of their time to preserve truth and extend reach. In the early centuries, people used parchment and scrolls to record revelation; Roman roads, built for imperial conquest, apostles used instead to carry the gospel to distant lands. Letters strengthened scattered believers, and physical gatherings in homes and public squares supported the oral tradition.

As history progressed, the tools evolved. The invention of the printing press in the fifteenth century fundamentally changed access to Scripture, moving the Word of God from the elite's domain into the hands of the common person. In later centuries, microphones amplified preaching to larger crowds, and radio and television extended the church's voice into living rooms across the globe.

Today, the internet, social media, and mobile devices have created a degree of connectivity that was once unimaginable.

None of these tools created the mission; they only served it. God gave us the mission, and the Holy Spirit gave us the power to carry it out. This same principle remains true today: technology may expand our reach, but it must never redefine our source, our authority, or our dependence. This distinction is critical because every generation is tempted to confuse usefulness with lordship. A tool that begins as a servant can quietly become a master when we stop asking what it is doing to our habits, our values, and our spiritual posture.

The issue is not whether the church should use tools—it always has. The issue is whether those tools remain submitted to Christ. When we surrender technology to divine purpose, it can help with proclamation, organization, and discipleship. *However, when it begins to shape our theology or replace our dependence on God, it becomes spiritually dangerous.*

God's mission is not merely a communication problem to be solved; it is a spiritual calling. Christ commissioned His followers to make disciples, preach the gospel, and bear witness to the kingdom (Matthew 28:19-20). Technology can support these efforts, but it cannot generate spiritual authority, replace the anointing of the Spirit, or breathe life into a dead heart.

The digital age intensifies ancient temptations—the desire to increase control, reduce waiting, and secure visible results. Yet the Spirit often leads through patience, surrender, and hiddenness. Technology accelerates distribution, but the Spirit forms depth. Therefore, the church must hold a mature posture: neither fearing tools irrationally nor trusting them uncritically.

The faithful path is Spirit-governed stewardship—using what serves the mission and rejecting what corrupts it. We must ensure that our innovation never outpaces our intercession.

MEDIA CAN AMPLIFY THE GOSPEL, BUT IT CANNOT REPLACE PRESENCE

The media defines the modern age. It has become the primary lens through which many people view the world and receive information. For the church, media offers unprecedented opportunities. Sermons stream live to thousands; Bible studies are archived for global access; worship music spreads across continents in minutes. These capabilities are real gifts when they remain under godly stewardship. They broaden access, preserve teaching, and let the message of Christ penetrate environments that were previously closed.

At the same time, the media has limits that the church must not ignore. It can transmit words and sounds, but it cannot fully replace the reality of embodied life together. Media is a flat representation of a multi-dimensional reality. It can show a baptism, but it cannot convey to a congregation the splash of water or the warmth of a physical embrace. It can broadcast a sermon, but it cannot look into the eyes of a grieving parishioner and offer the silent, steady presence of pastoral tenderness. Digital ministry is not the same as the incarnational life of the church; it remains partial.

Confusion grows when people begin to substitute access for involvement. Watching a service can easily replace gathering with the body; consuming spiritual content can be mistaken for actual spiritual formation. The Holy Spirit forms a people, not just an audience.

The New Testament emphasis on the "one another" commands—love one another, pray for one another, bear one another's burdens (Galatians 6:2)—requires physical proximity and shared life. Media can support the invitation to the community, but it cannot be the community.

Furthermore, media culture rewards traits that don't always align with spiritual maturity. It rewards brevity, novelty, and emotional intensity. It favors the spectacular over the ordinary and the polished over the authentic. If the church lets these media-driven values dictate its ministry, it may produce a version of faith that looks good but runs shallow.

The Spirit may lead a leader into a season of silence and slow formation, while digital culture demands a constant stream of "content." The church must remain vigilant:

Media must remain a means of distribution, never a master of the message.

ARTIFICIAL INTELLIGENCE AND THE TEMPTATION TO CONFUSE AUTOMATION WITH AUTHORITY

Artificial intelligence poses a sharper challenge than previous technological leaps because it appears to imitate thought rather than merely amplify action. Because AI can process vast amounts of data and produce persuasive, conversational language, it tempts the modern mind to treat it as a source rather than a servant. In some circles, people are increasingly fascinated with using AI to generate sermons, write prayers, or even offer spiritual counsel.

The church must maintain a clear theological distinction: AI can organize information, but it cannot produce revelation. It can summarize a doctrine, but it cannot anoint it. It can mimic the tone of devotion, but it cannot produce the fruit of devotion. These are the dividing lines between efficiency and life. The Spirit gives life; machines do not.

Revelation results from God speaking to man; automation results from man programming a machine. If we lose this distinction, we risk a future where the church is well-informed but spiritually hollow.

AI also intensifies the temptation to bypass dependence on the Holy Spirit. As human beings, we naturally prefer control over waiting and measurable output over prayerful uncertainty. AI offers instant responses. It makes productivity feel like wisdom and volume feel like fruitfulness. But speed is not truth, and a quickly generated result is not necessarily a word from God.

When a leader relies on an algorithm to do the spiritual labor of prayer and study, they trade anointing for automation.

Furthermore, AI is built on past patterns, meaning it reflects the biases and errors of its builders. It can produce confident errors and subtle distortions. Any use of these tools in ministry must remain under the authority of Scripture, under the guidance of a surrendered conscience, and under the Spirit's leadership.

The church must refuse the "efficiency trap" that suggests that because a machine can do it faster, it is doing it better. In the kingdom of God, the process matters as much as the product. The struggle of prayer, the labor of study, and the waiting on God are where the Spirit shapes the servant.

THE LIMITS OF DIGITAL SHEPHERDING

Shepherding is one of the most intimate metaphors for ministry. It implies a relationship between a shepherd who knows his sheep and sheep who recognize their shepherd's voice. In a technological age, we are tempted to "scale" shepherding through digital means.

While communication tools can help a pastor stay in touch with a large congregation, they have real limits.

when it comes to the deep care of souls.

Digital shepherding often prioritizes information delivery over relational presence. A machine or a pre-programmed system can produce comforting sentences, but it cannot intercede with the "groanings which cannot be uttered" (Romans 8:26). It cannot weep with those who weep or rejoice with those who rejoice in a way that truly shares the burden. It cannot discern spiritual warfare, identify hidden trauma, or confront a specific sin with the holy tenderness that only comes from a Spirit-led person.

The Holy Spirit leads people, not programs. Soul care is personal because God deals with us as persons, not data points. When believers begin treating technology as their primary counselor, they risk replacing communion with consumption. They may seek guidance without relationships and wisdom without accountability.

Digital tools can support shepherding at the margins—reminding people of truth or organizing practical care—but they must never become the center. The laying on of hands, the Spirit-led listening, and the intentional slowness required for faithful shepherding remain essential.

GUARDING AGAINST FLESH-DRIVEN MINISTRY IN A TECH AGE

The greatest danger to the church in a technological age is not opposition from the outside, but substitution from within. It is possible to build a ministry that is technologically advanced, strategically brilliant, and widely followed, yet spiritually dead. Flesh-driven ministry can appear strong and polished. It can move people emotionally and achieve impressive metrics. Yet, if it operates independently of prayer, humility, and Spirit-led discernment, it is "wood, hay, and stubble" (1 Corinthians 3:12).

Technology can easily mask a lack of anointing. A well-produced service can create an atmosphere of worship without God's presence. A clever social media post can garner likes without producing conviction.

When metrics become the new masters, leaders may begin to chase what performs rather than what transforms. They may feel pressure to remain relevant and visible, leading to a decrease in prayer and an increase in anxiety. This path leads to burnout and powerlessness.

The Spirit values integrity over optics. He values holiness over reach. The church must govern its content creation with discernment, recognizing that digital culture rewards constant reaction and noise.

Sometimes, the most Spirit-led thing a church can do is remain silent, go deep rather than wide, and refuse the urgency of the digital moment to stay aligned with God's eternal call. We must guard against the subtle belief that our success depends on our platform rather than our God.

KEEPING TOOLS IN THEIR PROPER PLACE

The church's framework for this age is simple but demanding: use tools as servants. Keep them in their proper place. Keep Scripture as the final judge, prayer as the first priority, and Christ as the absolute center.

We need not fear artificial intelligence as though it could rival God—it cannot. We need not fear media as though it could drown out the Spirit—it cannot. However, we must fear self-reliance, which is the surest pathway to spiritual decay.

Technology can extend our reach, help us organize for mission, and respond efficiently to the practical needs of our communities. It can bridge the unreached and serve the marginalized. But the Holy Spirit must remain the Lord over the process. He sustains our humility. He reminds us that our identity rests in Christ, not in our reputation or our digital influence.

The church's strategic advantage in a technological age is not its platform sophistication; it is the indwelling Spirit of God. He is the one who protects us from becoming efficient yet empty. He empowers our witness and ensures our message carries the fire of heaven.

God's mission will never advance by tools alone; it requires a people filled with the Spirit and surrendered to His lead. As we move forward, let us use every tool with wisdom, but let us trust only in the living God.

8

VATICAN II, ECUMENISM, AND THE QUESTION OF AUTHORITY

Institutional Power, Conscience, and the Spirit's Voice in a Technological Era

The twentieth century forced the church to confront rapid social, political, and technological change. Global wars, industrial expansion, mass communication, and shifting cultural norms altered the environment in which Christianity operated.

Within that context, the Second Vatican Council (1962–1965) convened with the stated aim of *a!iornamento*—bringing the church up to date—redefining the posture of the church in the modern world and addressing how it should engage contemporary realities.

Vatican II acknowledged that new forms of social communication were reshaping human interaction. Documents like *Gaudium et Spes* (The Pastoral Constitution on the Church in the Modern World) recognized that media, public opinion, and global awareness were transforming society, and the council acknowledged that, if used properly, these developments could spread the Kingdom of God.

This admission was significant; it marked institutional recognition that the church could not isolate itself from technological advancement. It had to engage it.

At the same time, the council's documents often reflected a cautious tone. While they affirmed the Spirit's presence in guiding the church, they placed heavy emphasis on institutional oversight and hierarchical authority.

In *Lumen Gentium* (The Dogmatic Constitution on the Church), the Holy Spirit is acknowledged as the one who "furnishes and directs [the Church] with various hierarchical and charismatic gifts." Yet, the council framed the administration of these gifts strictly within structured ecclesiastical governance. The tension between spontaneous spiritual empowerment and institutional control became evident.

The council also emphasized the dignity of the human person and religious freedom. It affirmed that individuals should not be coerced in matters of faith and that conscience plays a role in moral decision-making.

This language aligned with biblical themes of personal responsibility before God. At the same time, it raised enduring questions about how conscience interacts with institutional authority and whether structures might limit the Spirit's direct work in individuals.

The broader lesson emerging from this period is not confined to Roman Catholicism; it applies to all church traditions. As institutions adapt to modernity, they must decide how they exercise authority and honor the Spirit's freedom. Structures can protect doctrine and preserve order, but they can also become rigid if they resist renewal.

The Spirit is not hostile to order, but He is not confined by bureaucracy. In every generation, the church must examine whether its governance serves the Spirit's mission or restrains it.

That question has become even sharper in a technological era. Institutions now operate in a world of instant commentary, digital influence, global scrutiny, and competing voices that travel faster than formal deliberation.

Authority is no longer tested only in councils, denominations, or local assemblies. It is tested in public time, under pressure, before audiences that may never enter a sanctuary. In such an environment, the church must do more than preserve structure; it must prove that its authority remains alive, spiritually credible, and submitted to God's voice rather than merely protected by institutional habit.

INSTITUTIONAL AUTHORITY VERSUS SPIRIT-LED AUTHORITY

Authority in the church originates from Christ, who declared that all authority in heaven and earth had been given to Him (**Matthew 28:18**). That authority is mediated through the Spirit. In Acts, the Holy Spirit appoints overseers to shepherd the church of God (**Acts 20:28**). Leadership is not self-created; it is Spirit-recognized. The question, therefore, is not whether authority exists, but how it functions.

Institutional authority seeks stability and continuity. It establishes doctrinal boundaries, ordains leaders, and defines structure. These functions are necessary; without order, confusion spreads, and without accountability, abuse flourishes.

Yet, when institutional authority assumes exclusive control over spiritual legitimacy, tension arises. The Spirit distributes gifts according to His will (**1 Corinthians 12:11**). He may raise leaders from unexpected backgrounds or move outside traditional patterns.

The Pentecostal perspective often emphasizes immediacy, affirming that the Spirit empowers every believer, not only the clergy. Spiritual gifts are not restricted to hierarchy; lay members may exhibit prophecy, healing, and evangelistic boldness.

This democratization of spiritual empowerment challenges rigid structures and reflects Joel's prophecy that the Spirit would be poured out upon all flesh (**Joel 2:28**).

History shows that renewal movements often meet institutional resistance. Revival challenges routine, and spontaneous worship unsettles formal liturgy.

Institutions may fear disorder or theological drift. At the

same time, renewal movements must guard against arrogance and fragmentation; Spirit-led authority does not mean rejecting accountability. It means submission to Christ through the Spirit.

The balance lies in recognizing that the Spirit and structure are not enemies. The Spirit creates the church, and structure serves the church. The Spirit gives gifts, and order helps steward them. The problem begins when order claims final sovereignty.

No denomination, council, board, or hierarchy has the right to function as though it owns the Spirit's voice. Christ rules His church, and the Spirit remains the living executor of His will. Institutions are healthiest when they understand themselves as servants of that rule rather than substitutes for it.

This tension grows more visible in a connected age because authority is now constantly compared, challenged, clipped, broadcast, and debated. A believer can hear a bishop, a pastor, a theologian, an influencer, a critic, and a skeptic all before lunch. That reality can weaken healthy submission, but it can also expose shallow authority that depends too heavily on distance, secrecy, or inherited prestige.

The answer is not to abandon authority; the answer is to

restore its spiritual legitimacy. Authority must be biblical, accountable, humble, and visibly submitted to the Spirit.

ECUMENISM AND THE DESIRE FOR UNITY

Ecumenism emerged as a major theme of the twentieth-century church as leaders increasingly sought visible unity across denominational lines. This desire was driven in part by scandal; a fractured church preaching one gospel with many divided voices presents a weakened witness to the world.

Jesus prayed that His followers would be one (**John 17:21**), and we cannot dismiss that prayer lightly. Unity is not a minor aspiration; it is bound up with the credibility of Christian witness.

Vatican II contributed to this conversation by softening certain postures and opening greater dialogue. The council acknowledged that elements of sanctification and truth could be present outside formal Roman Catholic boundaries, referring to non-Catholic Christians as "separated brethren." That represented a shift in tone and a recognition that the Spirit's activity could not be discussed only in narrowly institutional terms.

Biblical unity, however, is never mere organizational merger. It is unity in truth, unity in Christ, and unity in the Spirit. Paul speaks of "one body, and one Spirit... one Lord, one faith, one baptism" (**Ephesians 4:4–5**). That kind of unity is spiritual before it is structural. It is deeper than cooperation and more costly than diplomacy; it requires humility, truthfulness, love, and submission to God. Without those elements, ecumenism can become a political arrangement with religious language rather than a true expression of Spirit-born unity.

Only the Spirit can sustain genuine Christian unity. Institutions can sign agreements, but no agreement automatically creates oneness. The Spirit must produce mutual love, doctrinal honesty, and the courage to resist both sectarian pride and careless compromise. Unity that ignores truth is brittle, and truth that refuses love is barren; the Spirit holds both together.

In a technological era that amplifies tribal loyalty and fragmentation, the church cannot afford to imitate the world's patterns. Ecumenical conversations must be anchored in prayer and a sincere desire for the kind of unity the Spirit creates.

CONSCIENCE, FREEDOM, AND THE INNER WITNESS OF THE SPIRIT

One of the most important themes raised by modern church discourse is the role of conscience. Conscience is not a replacement for Scripture, but it matters deeply because it is where the moral life becomes personal. Individuals do not stand before God merely as members of institutions; they stand before Him as persons who must answer for obedience and truthfulness.

Faith cannot be coerced. Institutions may teach, but they cannot manufacture inward conviction. The Spirit addresses the conscience directly, convicting of sin, righteousness, and judgment (**John 16:8**). He does not bypass the inner life in favor of external compliance. When institutions focus solely on conformity, they may produce outward order while failing to transform people.

This is not a call to radical individualism; conscience can be malformed or manipulated by culture. That is why Scripture must disciple conscience and the community must correct it. But even then, the church must make room for the Spirit's direct dealings with the soul.

The church must resist the error of idolizing the institution and treating conscience as dangerous, as well as the error of idolizing private judgment and treating every impression as divine authority.

The Spirit leads believers into truth, not isolation. He speaks personally, but never contradicts the revelation He has already given. In a technological era, algorithm-driven reinforcement can distort the inner life.

Many no longer know whether God has truly formed their convictions or whether their digital environment has merely amplified them. That is why the Spirit's inner witness matters so much; He purifies the conscience and teaches believers how to distinguish conviction from reaction and courage from ego.

THE SPIRIT'S VOICE IN A TECHNOLOGICAL ERA

The question of authority can no longer be discussed only in terms of councils and traditions. It must also be discussed in terms of media and the changing conditions under which people hear or receive authority.

The church now speaks into a world where people instantly compare every claim with others and where digital platforms reward certainty even when wisdom calls for patience.

The Spirit's voice is not silenced by technological change, but it can be crowded out. Public commentary can become so constant that believers lose the ability to listen, and institutional messaging can become so managed that spiritual candor disappears. The church must therefore protect spaces where prayer, silence, and holy listening remain possible.

In a technological era, authority must be transparent enough to be credible, humble enough to be approachable, and spiritual enough to be authoritative. The church does not need more bureaucracy; it needs more of the Spirit's presence.

When the Spirit is moving, authority is not something that must be constantly asserted; it is recognized because it carries the weight of truth and the fragrance of Christ.

Whether in formal councils or local assemblies, the goal remains the same: to remain a people governed by the Word and led by the Spirit, standing as a witness to the Kingdom of the Lord Jesus Christ, not the kingdom of this world.

9

THE HOLY SPIRIT AS THE CHURCH'S STRATEGIC ADVANTAGE

Why the Spirit Outperforms Systems, Strategy, and Artificial Intelligence

Jesus' promise to His disciples was not merely a promise of power, but a promise of perpetual guidance. He called the Holy Spirit the *Parakletos*—the one called alongside to help, counsel, and lead (John 14:16). This guidance is not a luxury for the spiritual elite; it is every believer's birthright and necessity. In a world of competing voices, complex choices, and rapid change, the Spirit serves as the supreme Guide who leads the church into all truth (**John 16:13**).

This guidance is profoundly personal. The Spirit does not merely provide a map; He provides His presence. He does not only point toward the truth; He is the Spirit of Truth. He leads through the Word, conscience, the community of faith, and the direct, inward witness of His Words. This leadership is active, intimate, and continuous. As Paul writes in Romans, "For as many as are led by the Spirit of God, they are the sons of God" (**Romans 8:14**). This responsiveness to the Spirit's lead evidences sonship.

In a technological age, data, algorithms, and predictive modeling are redefining the nature of guidance. We are trained to look to "the cloud" for answers, to trust analytics for direction, and to rely on automated recommendations for our next steps. While these tools offer practical utility, they cannot provide spiritual guidance. They can calculate patterns, but they cannot discern purpose. They can provide information, but they cannot provide revelation. The Holy Spirit remains the only Guide who knows God's mind and the human heart perfectly. His guidance is not based on historical data points but on eternal wisdom.

The church must therefore guard its attentiveness. If we become more practiced at consulting our devices than consulting the Spirit, our discernment will dull. If we prioritize the speed of a digital response over the "tarrying" required to hear from God, we will make efficient decisions that lack divine authority. The Spirit's guidance often requires silence, patience, and a willingness to be led in directions that do not make sense to the natural mind. To follow the Spirit is to acknowledge that His thoughts are higher than our thoughts and His ways higher than our ways (**Isaiah 55:8-9**).

DISCERNMENT IN AN AGE OF INFORMATION OVERLOAD

Discernment is the spiritual capacity to distinguish between truth and error, between the holy and the profane, and between the move of God and the impulses of the flesh. In the New Testament, discernment is presented as a gift of the Spirit and a discipline of the mature (**1**

Corinthians 12:10; Hebrews 5:14). It is the "spiritual sense" that allows the church to navigate a fallen world without being deceived by its patterns.

We now live in what is often called the "Information Age," but it might be more accurately described as the Age of Information Overload. We are inundated with a constant stream of content, opinions, narratives, and "truths." Digital platforms are designed to capture our attention and shape our perceptions. In this environment, the noise is so loud that the "still small voice" of the Spirit can easily be drowned out. Without active discernment, the believer is vulnerable to being "tossed to and fro and carried about with every wind of doctrine" (**Ephesians 4:14**).

Discernment is not merely about spotting what is obviously wrong; it is about perceiving what is subtly "off." It is the ability to recognize when a message is biblically accurate but spiritually hollow, or when a strategy is efficient but not Spirit-led. In a technological era, deception often comes dressed in the language of progress, empathy, or even spirituality. Artificial intelligence can now generate "spiritual" content that sounds remarkably like the real thing. Discernment is what allows the church to say, "This has the form of godliness, but denies the power thereof " (**2 Timothy 3:5**).

The Holy Spirit provides this discernment by anchoring us in the Word and sensitizing our spirits to Christ's character. He creates a "holy friction" when we encounter something that is not of God. This inward check is vital when navigating digital spaces where truth is often sacrificed for engagement and where the loudest voice is often mistaken for the most authoritative. The discerning believer learns to pause, to test the spirits (**1 John 4:1**), and to wait for the peace of God to act as an umpire in the heart (**Colossians 3:15**).

THE SPIRIT'S VOICE VERSUS THE ALGORITHMIC ECHO

One of the most significant spiritual challenges of the digital age is the "echo chamber" created by algorithms. Search engines and social media platforms are programmed to show us more of what we already like, believe, and interact with. This creates a feedback loop that reinforces our existing biases and isolates us from challenging truths. It rewards our preferences and punishes our attention if it wanders toward something that does not fit our profile. The Holy Spirit operates oppositely. He does not merely echo our preferences; He often challenges them. He does not only confirm what we already know; He leads us into "all truth," including the truths that are uncomfortable, convicting, and transformative. The Spirit is the Great Interrupter of our self-made echo chambers. He brings to mind the words of Jesus we might prefer to forget. He convicts us of the sins we have learned to justify. He pushes us toward the people we would rather avoid.

Algorithmic "guidance" is based on the past; the Spirit's guidance is based on the Father's will for the future. An algorithm can tell you what you are likely to do; the

Spirit tells you what you are called to do. To be Spirit-led is to be open to the divine "suddenly"—the moment when God breaks through our patterns and directs us toward a new work, a new sacrifice, or a new level of obedience.

The church must be careful not to let its own "ministry algorithms" replace the Spirit's voice. We can become so committed to our proven models, our successful patterns, and our demographic "hits" that we stop asking the Spirit for fresh direction. We can build churches that are perfectly optimized for a specific segment of the population while becoming deaf to the Spirit's call to reach the marginalized, the difficult, and the "lost." True guidance requires the courage to step out of the echo chamber and follow the voice that sounds like many waters (**Revelation 1:15**).

ARTIFICIAL INTELLIGENCE AND THE SEARCH FOR WISDOM

As artificial intelligence becomes more integrated into daily life, it is increasingly being positioned as a source of wisdom. We "ask" AI for advice on parenting, career choices, ethical dilemmas, and even biblical interpretation. While AI can synthesize vast amounts of human information and provide logical frameworks, it cannot possess or provide *Sophia*—the divine wisdom that comes from God.

Biblical wisdom is not merely applying information; it is "the fear of the Lord" (**Proverbs 9:10**). It is a moral and spiritual orientation that sees life from God's perspective. Wisdom is a gift the Spirit bestows on those who ask in faith (**James 1:5**). It involves discernment, timing, character, and an understanding of the heart. AI can provide a "smart" answer, but it cannot provide a "wise" one, because wisdom requires a soul, a conscience, and a relationship with the Creator.

The modern believer is tempted to settle for "smart" instead of "wise." We want the quick answer that solves the problem, rather than the deep wisdom that forms the person. We want the shortcut to a decision, rather than the process of seeking God. But the Spirit's wisdom often leads us down paths that look "foolish" to the world (1 Corinthians 1:25). His wisdom prioritizes the eternal over the temporary, the sacrificial over the selfish, and the glory of God over man's comfort.

In an age of AI, the church must double down on its commitment to spiritual wisdom. We must be a people who know how to wait on God until His wisdom settles the matter. We must recognize that the most "intelligent" solution is not always the "God-led" solution. Wisdom is found in the secret place, in meditating on the Word, and in the quiet whispers of the Spirit. No machine can replicate it, and no algorithm can produce it.

THE HOLY SPIRIT AS THE CORRECTIVE TO DIGITAL DECEPTION

The digital world breeds deception. From "deepfakes" to misinformation, from curated personas to manipulative marketing, the line between what is real and what is fabricated is constantly being blurred. This creates a culture of cynicism and distrust, where people no longer know what to believe.

The Holy Spirit is the supreme corrective to this deception. He is the Spirit of Truth who "searches all things, yea, the deep things of God" (**1 Corinthians 2:10**). He provides the believer with an "unction from the Holy One" so that they may know the truth (**1 John 2:20**). This internal "truth-meter" is essential for navigating a world of smoke and mirrors.

The Spirit's corrective work begins in the heart. He exposes our self-deceptions—how we pose, how we hide, and how we use technology to project an image that doesn't match our reality. He calls the church to authenticity in an era of curation. He reminds us that while man looks at the outward appearance (including the digital profile), God looks at the heart (**1 Samuel 16:7**).

Furthermore, the Spirit empowers the church to speak truth in a deceptive age. We are called to "speak the truth in love" (**Ephesians 4:15**). This means more than fact-checking; it means living out the gospel's truth in a way that exposes the world's hollow promises. In a culture of digital deception, a Spirit-filled life is the most powerful argument for the reality of God. Integrity, honesty, and transparency are not just moral choices; they are spiritual weapons.

REMAINING RESPONSIVE TO THE SUPREME GUIDE

The goal of the Christian life is not to become more efficient, but to become more like Christ. This requires a life that is fundamentally responsive to the Holy Spirit. We must learn to recognize His voice among the thousands of digital inputs. We must value His peace more than we value "engagement." We must follow His lead even when it contradicts the data.

This responsiveness is a habit we must cultivate. It requires intentional "disconnection" from the digital world to "connect" with the spiritual one. It requires the discipline of silence, the priority of prayer, and the humility to say, "Not my will, but Thine be done." As we move deeper into the technological age, the church must not become less dependent on the Spirit, but more so.

The Holy Spirit is not a resource to use; He is the Lord to follow. He is the supreme Guide who has navigated every cultural shift, every technological revolution, and every age of man with perfect wisdom. He is not confused by AI, overwhelmed by information, or hindered by algorithms. He is the same Spirit who hovered over the waters, spoke through the prophets, empowered the apostles, and dwells within us today. Our future does not belong to our tools; it belongs to our Guide. If we listen, discern, and obey, He will lead us through this age and into the fullness of all that God has planned.

10

A ROADMAP FOR THE 21ST CENTURY CHURCH

Spirit-Led Leadership, Innovation, and Discernment in the Age of AI

The story of the Holy Spirit does not merely fade into the background as history approaches its conclusion.

On the contrary, the Spirit who hovered over the primordial waters in Genesis is the same Spirit who, at the end of the age, joins the Bride in the final, urgent cry of the church: "And the Spirit and the bride say, Come" (Revelation 22:17).

This reveals a profound theological truth: the Spirit is the Alpha and the Omega of the church's experience. He began creation, initiated redemption in the incarnation, birthed the church at Pentecost, and will bring the mission to its glorious completion.

As the church navigates the complexities of the technological age, it must be steadied by the reality of the Spirit's finality. History is not a random series of technological disruptions or cultural shifts; it is a directed narrative overseen by the God who knows the end from the beginning (Isaiah 46:10). The Spirit ensures that God's purposes are not thwarted by human innovation, political upheaval, or spiritual opposition. He guarantees that what God has promised, He can also perform.

In a world obsessed with the "next big thing"—whether it be the next iteration of artificial intelligence, the next digital frontier, or the next social revolution—the Spirit reminds the church of the "final thing." He directs our gaze to Christ's return and the restoration of all things.

This eschatological perspective is not a call to escape from the world, but a motivation to live faithfully within it. Because the Spirit will finish what He started, we can serve with confidence that the pace of change cannot shake us. We are not working toward a precarious future; we are working from a secured victory.

The Spirit's role, in the end, also involves preparing the church. He sanctifies the Bride, presenting her to Christ "not having spot, or wrinkle, or any such thing; but that it should be holy and without blemish" (Ephesians 5:27).

This refining work is often intense, pruning self-reliance and exposing false substitutes. In an age of automation, the Spirit continues this deep, manual labor of character formation, ensuring that the church's witness remains pure and its identity remains anchored in Christ alone.

FAITHFULNESS OVER FASCINATION: THE CALL TO DISCERNMENT

As we look toward the future, the church is often tempted by two extremes: a fearful retreat from technology or an uncritical fascination with it. Both stem from a lack of trust in the Holy Spirit.

Fear suggests the Spirit is not powerful enough to protect the church from digital influence; fascination suggests technology is a greater source of progress than the Spirit's power.

The call at the end of this age is not for fascination, but for faithfulness. Faithfulness requires us to use the tools of our time without bowing to them. It requires us to innovate without losing our reverence for the ancient paths. It requires us to engage the modern world without absorbing its self-reliance.

As theologians such as Jürgen Moltmann note, the Spirit is the "Spirit of life" who breaks into our closed systems of technology and death, offering a future that is not merely a projection of our present capabilities, but a gift from God.

This faithfulness is expressed through discernment. As we have explored throughout this work, discernment is the ability to see beneath the surface of innovation to the spiritual realities beneath. It is the wisdom to know when a digital shortcut is actually a spiritual detour.

It is the courage to choose "slow" discipleship over "fast" consumption.

The Spirit-led church does not ask, "Can we do this?" as its primary question. It asks, "Should we do this? Does this lead to holiness? Does this glorify Christ? Does this maintain our dependence on God?"

The pressures of the age are real, but they are not final. The pace of change is intense, but it has not weakened God's hand. The church must move forward with open eyes and a guarded heart. We must recognize that the most technologically advanced congregation is not necessarily the most Spirit-led.

The most efficient platform is not necessarily the most anointed. We measure faithfulness by our alignment with the Spirit's voice, not by our alignment with digital trend cycles.

THE SPIRIT AS THE CORRECTIVE TO SELF-RELIANCE

At its core, the technological age is an age of self-reliance. We have built systems that give us the illusion of omniscience (through the internet), omnipresence (through social media), and omnipotence (through automation).

We have become the architects of our own environments, the curators of our own truths, and the masters of our own schedules. This environment makes the biblical call to "lean not unto thine own understanding" (Proverbs 3:5) feel increasingly counter-cultural.

The Holy Spirit serves as the ultimate corrective to this self-reliance. He reminds the church that "it is not in man that walketh to direct his steps" (Jeremiah 10:23). He continually calls us back to prayer, waiting, and surrender.

He exposes the fragility of our systems and the limits of our intelligence. He reminds us that while we can build a tower that reaches toward the heavens, only He can bring the fire of heaven down to earth.

This corrective work is essential for the church's survival. If we try to finish in the flesh what was begun in the Spirit (Galatians 3:3), we will produce a ministry that is efficient but empty. We will build institutions that look successful on a dashboard but are bankrupt in the presence of God.

The Spirit will finish what He started, but He will not be drafted into our self-made plans. He requires us to lay down our tools at His feet and ask for His breath.

The "future" of the church, therefore, is not a technological problem to be solved; it is a spiritual posture to be maintained.

Our strategic advantage in a world of AI and automation is our capacity to be led by the Spirit of the living God.

When the world is overwhelmed by information, we must be grounded in revelation. When the world is driven by speed, we must be anchored in peace.

When echo chambers fracture the world, we must be unified by the Spirit.

THE UNFINISHED MISSION AND THE FINAL OUTPOURING

The mission of the church remains unfinished. Billions remain unreached, and the complexities of the modern world have created new frontiers of spiritual need. Yet, we do not face this unfinished task with diminished power.

The same Spirit who descended at Pentecost is still moving with global authority. He is opening doors in closed nations, awakening hunger in secular cities, and empowering a new generation of witnesses.

Scripture suggests that as the end approaches, the Spirit's work will intensify. The "latter rain" of the Spirit is promised to be even more significant than the former (Joel 2:23; James 5:7). This should steady the church.

We are not presiding over a dying movement; we are participating in a surging one. The Spirit is currently coordinating a global symphony of prayer, witness, and revival that transcends denominational lines and geographic boundaries.

Our role is to remain attentive to this move. We must not be so distracted by the "noise" of the technological age that we miss the "signal" of the Spirit's activity. He may be moving in ways that do not fit our traditional categories or our digital metrics.

He may be using people we would overlook and methods we would find unconventional. But wherever Christ is exalted, wherever the Word is honored, and wherever hearts are surrendered, the Spirit is at work.

The mission will be finished not because we have perfected our systems, but because the Spirit has not failed. He will ensure that "this gospel of the kingdom shall be preached in all the world for a witness unto all nations; and then shall the end come" (Matthew 24:14).

Our tools may assist in this distribution, but the Spirit will supply the conviction. He will bring the harvest. He will finish the story.

STRENGTH FOR THE TRANSITION

We are living in a period of profound transition. The shift from the industrial age to the technological age has changed how we think, relate, and believe. For many in the church, this transition is disorienting. It feels as though the foundations are being shaken and the ancient ways are being forgotten.

In this transition, the Spirit provides "strength for the journey." He is our Comforter in the literal sense—the one who provides strength. He reminds us that while the "grass withereth, the flower fadeth: but the word of our God shall stand for ever" (Isaiah 40:8).

He provides the continuity that our souls crave. He connects us to the saints of old and the church of the future, reminding us that we are part of an "unshakable kingdom" (Hebrews 12:28).

The Spirit also provides the wisdom needed for new frontiers. He is not "catching up" to technology; He is ahead of it. He knows how to use AI for the gospel without letting it corrupt the gospel. He knows how to shepherd souls in a digital world without losing the personal touch.

He provides the "anointing" that breaks the yoke of modern addiction, modern distraction, and modern despair. We do not need to be experts in every new technology; we need to be experts in the voice of the Spirit.

The church must move forward with "sober mindedness." We should not be alarmists, nor should we be enthusiasts. We should be a people of the Spirit. We should use tools with wisdom, but we must place our trust in nothing but the living God.

As history accelerates, our roots must go deeper into the Spirit. Our prayer life must be more robust, our discernment sharper, and our surrender more complete.

A FINAL CALL TO SURRENDER

The end of this book is not a conclusion, but an invitation. It invites us to reclaim the Holy Spirit as the central reality of the church's life. It is an invitation to stop treating the Spirit as a doctrine to be affirmed and start treating Him as a Person to be obeyed.

It invites us to lay down our technological self-reliance and return to the upper room.

The Spirit who began this work has not lost His authority over history. He started this story in Genesis; He intensified it at Pentecost; and He will finish it in glory.

The question for the modern church is simple: will we go through the motions of ministry, or will the breath of God animate us? Will we build monuments to our own innovation, or will we be a temple for the Holy Spirit?

The future does not belong to the most technologically advanced congregation. It does not belong to the loudest voice, the fastest platform, or the most efficient machine.

It belongs to the Spirit-led church—the praying church, the discerning church, the holy church, the church that still knows how to wait on God and obey when He speaks.

The Spirit and the Bride say, "Come." The Spirit who moved over the waters is moving now. He is calling the church to deeper dependence, clearer discernment, and greater holiness. He is ready to finish what He started.

Our task is to say "Yes." Our task is to surrender. Our task is to be the people through whom the Spirit speaks to a technological world that is desperately hungry for life.

With the Spirit, life cannot be extinguished. Without the Spirit, there is activity but no life. He started this story, and He will finish it. Even so, come, Lord Jesus. The grace of our Lord Jesus Christ be with you all. Amen. (Revelation 22:20-21).

THE SPIRIT WILL FINISH WHAT HE STARTED

Unchanging Power, Unfinished Mission, and Hope Beyond Technology

The story of the Holy Spirit is not a chain of isolated moments scattered across Scripture and history. It is one continuous movement from creation to consummation. The Spirit who hovered over the waters in Genesis, who empowered judges and prophets, who rested upon Christ, who descended at Pentecost, and who ignited revival across continents is still active now. He has not diminished. He has not withdrawn. Progress has not displaced Him, innovation has not outrun Him, and the rise of digital systems has not made Him unnecessary. The same Spirit who moved in the beginning is moving still.

From the beginning, the Spirit has been God's executive presence in the world. He creates. He sustains. He convicts. He empowers. He sanctifies. He sends. He strengthens. He finishes what He begins. The Old Testament revealed His cosmic authority, His empowering presence among God's people, and His inward transforming work. The New Testament unveiled His indwelling presence, His universal outpouring, and His active leadership in the life of the church. Christian history has testified again and again that when the Spirit moves, the church comes alive. And the technological age has not interrupted that trajectory. It has only made the

church's need for Him more obvious.
That need is urgent because the modern world offers endless substitutes for dependence. It offers speed in place of waiting, visibility in place of depth, output in place of fruit, information in place of revelation, and efficiency in place of spiritual power. The church can now distribute content more widely than at any other point in history. It can automate tasks, archive teaching, analyze behavior, and coordinate ministries across vast distances. These developments are real, and many can serve God's mission. But none of them can replace the Spirit. Technology can extend reach. Only the Spirit can extend life. Tools can assist ministry. Only the Spirit can animate it. Systems can organize activity. Only the Spirit can create holy people.

This is where the church must remain clear-eyed. Artificial intelligence, like every other tool, presents both opportunity and temptation. It can help translate biblical resources, reduce administrative strain, and make practical labor more efficient. Yet it also tempts leaders to confuse automation with authority, speed with wisdom, and convenience with guidance. It can imitate thought, but it cannot produce revelation. It can summarize truth, but it cannot apply truth to the conscience with divine power. It can generate language, but it cannot generate the fear of the Lord. The Spirit is still the Spirit of truth, and the church must keep every tool beneath Scripture, beneath conscience, and beneath His leadership.

That means the future of the church will not be secured

by panic over technology, nor by naive surrender to it. Fidelity to the Spirit will secure it. Prayer must remain central.

Scripture must remain supreme. Holiness must remain essential. Discernment must remain active. The church must not become intoxicated with its own machinery. It must not confuse polished ministry with powerful ministry. It must not measure success only by what can be counted in public while ignoring what God is shaping in secret. A church may gain followers and still lose its fire. It may refine its systems and still neglect surrender. But where the Spirit remains honored, even simple ministry can carry weight that no human system can manufacture.

The church's great hope is that it does not carry this mission in its own strength. The Spirit still calls leaders. He still convicts sinners. He still regenerates hearts. He still empowers witness. He still breaks pride. He still heals what sin has shattered. He still crosses barriers that culture, language, politics, and technology cannot fully overcome. He still breathes on ordinary obedience and turns it into something eternal. He still forms communities of truth, holiness, courage, and love in a distracted, accelerating world.

This should steady the church. The future does not belong to the most technologically advanced congregation. It does not belong to the loudest voice, the fastest platform, or the most efficient machine. It belongs to the Spirit-led church, the praying church, the discerning church, the holy church, the church that still knows how to wait on God and obey when He speaks. The mission remains unfinished, but the power has not failed. The pressures of the age are real, but they are not final. The pace of change is intense, but it has not weakened God's hand. The Spirit who began this work has not lost His authority over history. He will guide the church through every transition, challenge, false substitute, and new frontier that emerges.

So, the call at the end is not fear, but faithfulness. Not retreat, but discernment. Not fascination with the future, but surrender to the God who already stands in it. The church must move forward with open eyes and a guarded heart. It must use tools without bowing to them. It must innovate without losing reverence. It must engage the modern world without absorbing its self-reliance. It must remember what has been true from Genesis until now: without the Spirit, there is activity but no life. With the Spirit, life cannot be extinguished. He started this story, and He will finish it.

REFERENCES & WORKS CITED

Ministry-Friendly, Simplified Citation Style

(All Scripture quotations from the King James Version unless otherwise noted.)

SCRIPTURE REFERENCES

Genesis 1:2 — "And the Spirit of God moved upon the face of the waters."
Genesis 2:7 — "Breathed into his nostrils the breath of life."
Psalm 33:6 — "By the word of the Lord were the heavens made..."
Job 33:4 — "The Spirit of God hath made me..."
Psalm 104:30 — "Thou sendest forth thy spirit, they are created..."
Psalm 51:10–12 — David's prayer for renewal.
Ezekiel 36:26 — "A new heart also will I give you..."
Joel 2:28 — "I will pour out my spirit upon all flesh..."
Jeremiah 29:11 — God's promise to the exiles.
John 3:5–6 — Born of water and Spirit.
John 14:16, 26 — Promise of the Comforter.
John 16:7–15 — The Spirit of truth guiding believers.
Acts 1:8; 2:1–4; 2:41; 2:47 — Pentecost and early church growth.
Romans 8 — Life in the Spirit.
1 Corinthians 3:6 — "God gave the increase."
Galatians 5:22–23 — Fruit of the Spirit.
Revelation 22:17, 20–21 — "The Spirit and the bride say, Come."

THEOLOGIANS & SCHOLARS

Fee, Gordon. Referenced for Pauline pneumatology and the Spirit as the personal presence of God.
Quoted in manuscript: "The Spirit's role in the New Covenant is the fulfillment of this life-giving presence."
Chan, Simon. Cited regarding ecclesial identity and Pentecostal theology.
Quoted in manuscript: "True Pentecostal growth is marked by this 'ecclesial identity.'"
Moltmann, Jürgen. Referenced concerning the Spirit as the "Spirit of life."
Quoted in manuscript: "The Spirit is the 'Spirit of life'..."

CHURCH DOCUMENTS & HISTORICAL SOURCES

Vatican II — *Gaudium et Spes*.
Referenced regarding human progress and the Spirit guiding modern engagement.
Quoted in manuscript: "Human progress... must be governed by the Spirit to serve the true good of humanity."
Vatican II — *Lumen Gentium*.
Referenced concerning charismatic and hierarchical gifts.
Quoted in manuscript: "The Holy Spirit... furnishes and directs the Church..."
Azusa Street Revival (1906).
Referenced as the catalyst of modern Pentecostalism.
Quoted in manuscript: "The Azusa Street revival... became a catalytic event."
Parham, Charles Fox.
Referenced for early Pentecostal teaching on Spirit

baptism.
Quoted in manuscript: "Parham taught that speaking in other tongues was a sign..."

Bennett, Dennis.
Referenced for the Charismatic renewal in mainline churches.
Quoted in manuscript: "Dennis Bennett's public testimony... became a turning point."

MODERN THEMES & CONCEPTS

Artificial Intelligence & Automation.
Discussed as tools that cannot replace spiritual authority.
Example: "AI can organize information, but it cannot produce revelation."

Digital Media & Technology.
Addressed as helpful but spiritually insufficient.
Example: "Media can amplify the gospel, but it cannot replace presence."

Global Pentecostal Growth.
Referenced as evidence of Spirit-driven expansion.
Example: "Africa... witnessed exponential growth throughout the twentieth century."

So the call at the end is not fear, but faithfulness. Not retreat, but discernment. Not fascination with the future, but surrender to the God who already stands in it. The church must move forward with open eyes and a guarded heart. It must use tools without bowing to them. It must innovate without losing reverence.

About the Author

Bishop Kevin Eugene Knight, Sr.

Jurisdictional Bishop, New Jersey Third Ecclesiastical Jurisdiction; Pastor, Heavenly Temple Church of God in Christ, Jersey City, NJ

Bishop Kevin Eugene Knight, Sr. is an observed leader whose life reflects steady growth in faith, service, and visionary leadership. Born in 1954 in Jersey City, New Jersey, he is the fourth of seven children blessed to the late Bishop John L. and Mother Fannie B. Knight. Raised in a home grounded in faith and community responsibility, he embraced salvation in 1975 and accepted his call to ministry in 1976. In 1977, he married Mary L. Knight, the love of his youth. Their union has been blessed with two children—Kevin Eugene Jr. and Shantae' Merie (William Gary Young)—and treasured grandchildren.

Bishop Knight began his academic preparation at St. Peter's College in Jersey City, where he earned a Bachelor of Science in Urban Studies and an Associate of Arts in Public Policy. He later pursued advanced theological training at Grace Hill Bible University, earning both the Master of Divinity and Doctor of Theology degrees.

His ministry journey reflects consistent growth and increasing responsibility. Beginning in the mid-1990s, he served as District Secretary and District Treasurer,

President of the Jurisdictional Youth Department, and Chairman of AIM. His administrative gifts led to his later appointment as Jurisdictional Secretary. In 1999, he became Pastor of Heavenly Temple Church of God in Christ, where he continues to serve faithfully.

His national ministry began when he was appointed Overseer in the COGIC Adjutancy Corps. Bishop Knight later served as Vice President of the International Youth Department, appointed by Presiding Bishop G. E. Patterson, and contributed significantly to the AIM Advisory Board. His commitment to excellence in church governance led to his tenure on the General Assembly Resolution Committee, where he eventually served as Chairman.

He was elevated to a higher ecclesiastical office in August 2016, when Bishop Jerry W. Macklin consecrated him (on-site) to the sacred office of Bishop. In November of that same year, the Chief Apostle of the Church of God in Christ, Bishop Charles Edward Blake, Bishop Knight received a "double-portion" through the "laying on of hands." In April 2017, he was ratified as Jurisdictional Bishop of the New Jersey Third Ecclesiastical Jurisdiction. In 2024, he was elected Associate Justice of the Judiciary Board of the Church of God in Christ, Inc., further affirming his reputation for wisdom, fairness, and integrity.

Bishop Knight's influence extends deeply into the community. Since 2002, he has served as President and Executive Director of the Community Outreach Team, Inc. (COT), a faith-based nonprofit his father, Bishop

John L. Knight, originally envisioned. Under Bishop Kevin Knight's leadership, COT is advancing a transformative $90 million revitalization project that will include over 100 units of senior housing and a recreational facility —an enduring testament to his commitment to community uplift.

He also serves on New Jersey Together, an ecumenical coalition advocating for justice and equity, and previously served on the Pontotoc Board of the Church of God in Christ, Inc.

Throughout his life, Bishop Knight has been celebrated for his administrative excellence, strategic insight, and ability to unify people toward shared goals. His leadership is marked by humility, compassion, and a deep sense of responsibility to God and community. His foundational scripture, *"The steps of a good man are ordered by the Lord"* (Ps. 37:23), continues to guide his journey.

Bishop Knight remains a devoted husband, father, grandfather, pastor, and servant of God—steadfast in his desire to one day hear the words, "Well done, thy good and faithful servant."